Bioenergetics & Metabolism of Green Algae: II

Papers by
Marcia Brody, George Papageorgiou, G. P. Findlay et al.

RA recent articles & research in progress RIP

IN COOPERATION WITH THE
SMITHSONIAN SCIENCE INFORMATION EXCHANGE

Summaries of current research projects are included in the final section of this volume. Previously unpublished, these summaries were obtained from a search conducted by the Smithsonian Science Information Exchange, a national collection of information on ongoing and recently terminated research.

MSS Information Corporation
655 Madison Avenue, New York, N.Y. 10021

Library of Congress Cataloging in Publication Data
Main entry under title:

Bioenergetics and metabolism of green algae.

Vol. 2: Papers by Marcia Brody, George Papgeorgiou, G. P. Findlay, et al.
1. Chlorophyceae--Addresses, essays, lectures.
2. Bioenergetics--Addresses, essays, lectures.
3. Plants--Metabolism--Addresses, essays, lectures.
I. Cárdenas, J. [DNLM: 1. Algae, Green--Collected works. 2. Algae, Green--Metabolism--Collected works.
QK569.C6 B615 1974]
QK569.C6B56 589'.47'0419121 74-515
ISBN 0-8422-7200-3

TABLE OF CONTENTS

CREDITS AND ACKNOWLEDGEMENTS

Bishop, Norman I.; and James Wong, "Observations on Photosystem II Mutants of Scenedesmus: Pigments and Proteinaceous Components of the Chloroplasts," *Biochimica et Biophysica Acta*, 1971, 234:433-445.

Brody, Marcia, "The Influence of Two Naturally Occurring Factors, from the Leaves of *Ricinus communis*, on the Fluorescence Induction of Higher Plant Chloroplasts and Intact Algae," *Biophysical Journal*, 1971, 11:189-203.

Brody, Marcia; and James Edward White, "Environmental Regulation of Enzymes in the Microbodies and Mitochondria of Dark-Grown, Greening, and Light-Grown *Euglena gracilis*," *Developmental Biology*, 1973, 31:348-361.

Delrieu, M.-J.; and Y. de Kouchkovsky, "Relationships between the Photon Distribution between the Two Photosystems, the Concentration of System II Reaction Centers and the Intersystem Equilibrium Constant in *Chlorella pyrenoidosa*," *Biochimica et Biophysica Acta*, 1971, 226:409-421.

Duysens, L.N.M., "3-(3,4-Dichlorophenyl)-1,1-Dimethylurea (DCMU) Inhibition of System II and Light-Induced Regulatory Changes in Energy Transfer Efficiency," *Biophysical Journal*, 1972, 12:858-863.

Findlay, G.P.; A.B. Hope; and N.A. Walker, "Quantization of a Flux Ratio in Charophytes?," *Biochimica et Biophysica Acta*, 1971, 233: 155-162.

Goodenough, Ursula W.; and L. Andrew Staehelin, "Structural Differentiation of Stacked and Unstacked Chloroplast Membranes: Freeze-Etch Electron Microscopy of Wild-Type and Mutant Strains of *Chlamydomonas*," *The Journal of Cell Biology*, 1971, 48:594-619.

Matucha, M.; L. Žilka; and K. Švihel, "Gas Chromatographic Analysis of the Higher Fatty Acids of the Alga *Chlorella vulgaris (pyrenoidosa)*," *Journal of Chromatography*, 1972, 65:371-376.

Mosser, Jerry L.; Nicholas S. Fisher; and Charles F. Wurster, "Polychlorinated Biphenyls and DDT Alter Species Composition in Mixed Cultures of Algae," *Science*, 1972, 176:533-535.

Papageorgiou, George; and Govindjee, "pH Control of the Chlorophyll *a* Fluorescence in Algae," *Biochimica et Biophysica Acta*, 1971, 234: 428-432.

Sirevåg, Reidun; and Robert Paul Levine, "Fatty Acid Synthetase from *Chlamydomonas reinhardi*: Sites of Transcription and Translation," *The Journal of Biological Chemistry*, 1972, 247:2586-2591.

White, R.C.; and G.A. Barber, "An Acidic Polysaccharide from the Cell

Wall of *Chlorella pyrenoidosa,*" *Biochimica et Biophysica Acta,* 1972, 264:117-128.

Zumft, Walter G.; and Hart Spiller, "Characterization of a Flavodoxin from the Green Alga *Chlorella,*" *Biochemical and Biophysical Research Communications,* 1971, 45:112-118.

PREFACE

The green algae are the most primitive eukaryotic organisms of the plant kingdom. They are of enormous ecological importance as major fresh water nitrate users. Their bioenergetic metabolism has been intensely studied, and much has been learned about the basic photochemical processes underlying photosynthesis.

The present two-volume collection of papers provides a comprehensive survey of current research on the important enzymatic processes associated with photosynthesis, as well as the biochemistry of dark respiration. Many of the papers in this collection deal with the chemistry and function of the algal chlorophylls and the various processes of electron transfer associated with photosynthesis. Other reports concern fatty acid and polysaccharide synthesis in green algae, and the effects of insecticides on algal metabolism. A further topic of the papers included is chloroplast structure, complementing the intensive research into chloroplast function. The volumes form part of MSS' continuing series of current papers on subjects of major ecological interest.

Photosynthetic Pigments and Photosynthetic Electron Transfer Components

Environmental Regulation of Enzymes in the Microbodies and Mitochondria of Dark-Grown, Greening, and Light-Grown *Euglena Gracilis*

MARCIA BRODY[1] AND JAMES EDWARD WHITE[2]

Catalase activity is demonstrated histochemically in the microbodies of aerated cultures of *Euglena gracilis* strain Z grown on inorganic media supplemented with acetate or glucose. Although this enzyme can also be assayed photometrically in cell-free extracts of acetate-supplemented cells, it is below the level of detectability in extracts of glucose-supplemented cells, there being an order of magnitude fewer microbodies in the latter than the former. Even acetate-supplemented cultures (dark-grown, greening, or continuously light-grown) fail to exhibit detectable catalase activity when CO_2 is removed from the air by Ascarite.

Negative results were obtained with histochemical techniques considered optimal for the demonstration of cytochrome oxidase; under other conditions, however, a KCN-sensitive enzyme was revealed in the mitochondrial matrix. This (unidentified) enzyme is first observed in mitochondria after 20–24 hr of greening, reaches a maximum intensity at about 48 hr, and becomes undetectable by 72 hr of greening. Poisoning of photosynthesis by 3-(3,4-dichlorophenyl)-1,1-dimethylurea (DCMU) results in loss of activity of this mitochondrial enzyme.

INTRODUCTION

The term "microbody" was first introduced to describe the single membrane-bound organelles (approximately 1 μm in diameter) observed in kidney and liver cells by means of electron microscopy

[1] This research was partially supported by a Faculty Research Grant to M. B. from The City University of New York.

[2] Work to be submitted in partial fulfillment of the requirements for the Ph.D. degree at The City University of New York.

(Rhodin, 1954; Gänsler and Rouiller, 1956). Similar structures have been found in higher plants (Mollenhauer *et al.*, 1966; Breidenbach and Beevers, 1967; Breidenbach *et al.*, 1968; Frederick *et al.*, 1968; Tolbert and Yamazaki, 1969) and microorganisms (Müller, 1969; Graves *et al.*, 1971a; Gerhardt and Berger, 1971; Hanzely *et al.*, 1971; Gergis, 1971). The biological significance of catalase in microbodies was recognized by de Duve (1965) and de Duve and Baudhuin (1966). Subdivisions of microbodies into glyoxysomes and peroxisomes was made on the basis of glyoxysomes containing the enzymes involved in the glyoxylate path or cycle (Breidenbach and Beevers, 1967) and peroxisomes containing the enzymes of the glycolate cycle (Frederick and Newcomb, 1969), which in the case of leaf peroxisomes, function in photorespiration (Tolbert *et al.*, 1968; Kisaki and Tolbert, 1969). Not only has it become evident that microbodies are heterogeneous in their enzymatic composition, but also that functions ascribed exclusively to these organelles may indeed be shared by others. Müller (1969) and Hogg (1969) have shown that in *Tetrahymena pyriformis* three enzymes of the glyoxylate cycle are found inthe mitochondriaand two (bypass) in the microbodies.

Although in their electron microscopic studies of *Euglena gracilis* var. *bacillaris*, Graves *et al.* (1971a) and Hanzely *et al.* (1971) demonstrated the existence of microbodies, they reported being unable to detect catalase in these organelles. In the present work, catalase activity is observed with aerated cultures of *Euglena gracilis* strain Z: (a) in microbodies of cells grown on inorganic media supplemented with acetate or glucose by the enzymatic oxidative polymerization of 3,3′-diaminobenzidine (DAB) in the presence of hydrogen peroxide (Vigil, 1970), and (b) in cell-free extracts of acetate-supplemented cells by Lück's (1963) spectrophotometric assay; catalase activity is not discernible with cells grown in CO_2-free air.

In the present work a mitochondrial enzyme was disclosed in the matrix, under conditions of DAB-incubation other than those considered optimal (Novikoff and Goldfischer, 1969; Gerhardt and Berger, 1971) for cytochrome oxidase; the activity of this enzyme is light-induced, transitory, and possibly photosynthesis dependent.

MATERIALS AND METHODS

Cultures

Growth in air. Euglena gracilis Klebs strain Z Pringsheim was obtained from Dr. S. H. Hutner of the Haskins Laboratory, New York City, New York. Three types of light regime were used in growing these cells. "Light-grown" cells were cultured continuously under conditions in which the incident light (incandescent) falling on the culture flasks was approximately 1.9×10^3 erg cm^{-2} sec^{-1} (about 110 ft-c) as determined with a Yellow Springs Radiometer, Precision Thermistor Model 65 (Yellow Springs Instrument Co., Yellow Springs, Ohio). "Dark-grown" cells were subcultured from cells depleted of their chlorophyll and chloroplasts by being grown in aluminum foil wrapped flasks, in a darkroom, for more than 30 generations. During this depletion process, fresh cultures were maintained by periodic transfer while the cells were in the log phase of growth, i.e., about every 4 days. We use the term "greening" to refer to dark-grown cells placed in the light (as above) for indicated periods of time. Vigorous agitation with a

Burrell wrist-action shaker (Burrell Corp., Philadelphia, Pennsylvania) was used to aerate the three types of cultures maintained at 23–26°C. Cells were grown in cotton-stoppered, 250-ml Erlenmeyer flasks, containing 100-ml aliquots of the medium described by Cramer and Myers (1952) supplemented with either acetate or glucose.

Growth in CO_2-free air. For these experiments, cultures were aerated by bubbling with air that had passed through a train of U-tubes filled with excess Ascarite (Arthur H. Thomas Co., Philadelphia, Pennsylvania). Since preliminary experiments had indicated that growth in CO_2-free air was severely retarded, large inocula (10^8 cells) in the exponential phase of growth, were used. Light-grown cells were exposed for 72 hours to both light and CO_2-free air, while dark-grown cells were kept in the dark during the same interval of CO_2-free growth; aliquots of such dark-grown cells were allowed to green in CO_2-free air and light for periods of 12, 24, or 48 hours.

Inhibition of Photosynthesis

Photosynthesis was inhibited by adding 3-(3,4-dichlorophenyl)-1,1-dimethylurea (DCMU) (Sigma Chemical Co., St. Louis, Missouri) which was dissolved in ethanol (to give a stock solution of 10^{-2} *M*) and added to culture flasks to yield a final concentration of 1×10^{-5} *M*.

Cell-Free Extracts

For the preparation of cell-free extracts to be used for Lück's (1963) photometric assay for catalase, cells were harvested by centrifugation at 500 *g* for 5 min, and then washed once with, and resuspended in, 0.4 *M* sucrose in 50 m*M* potassium phosphate buffer, pH 7.0, to give a 20% (v/v) cell suspension. The cells in suspension were broken by sonication in a Biosonik III (Bronwill Scientific Co.,

Rochester, New York) for 15 sec at a frequency of 20 kilocycles/sec. The suspension of completely broken cells (light microscopy) was centrifuged at 250 *g* for 10 min to remove whole cells, chloroplasts and large chloroplast fragments. The cell-free extracts which we utilized in Lück's (1963) photometric assay, correspond to 2.0×10^6 cells/ml. Protein content per milliliter was determined, according to Lowry *et al.* (1951), by using a calibration curve obtained with crystalline bovine serum albumin (Sigma Chemical Co., St. Louis, Missouri). Our results are given in "units" of catalase per milligram protein, the definition of "unit" being that amount of enzyme which decomposes half the H_2O_2 present in 100 sec at 25°C (Lück, 1963).

Electron Microscopy

Pellets of *Euglena* cells were obtained for electron microscopy by gentle centrifugation at 500 *g* for 5 min, and fixed for 2 hr at 4°C with 2.5% glutaraldehyde (Polyscience Inc., Rydel, Pennsylvania) in 0.05 *M* cacodylate buffer (sodium cacodylate; Matheson, Coleman and Bell, Norwood, Ohio) at pH 7.4 (Vigil, 1970). The cells were then washed several times at room temperature with 0.1 *M* cacodylate buffer and either used for cytochemistry (see below) or immediately placed in 2% osmium tetroxide in 0.05 *M* cacodylate buffer, pH 7.4, for 2 hr at 4°C. Cells were dehydrated in ethanol, passed through propylene oxide, and then embedded in Epon 812. Thin sections (approximately 500 Å) were cut on a Sorvall Porter-Blum MT-2 microtome with a glass or diamond knife, stained with uranyl acetate (Watson, 1958) followed by lead citrate (Coggeshall, 1965), and viewed in an RCA EMU 3-H electron microscope.

For cytochemical studies, glutaraldehyde-fixed cells were washed several times at room temperature in 0.1 *M* cacodylate buffer and equilibrated in two washings of 0.1 *M* AMP buffer (2-amino-2-methyl-1,3-propandiol; Sigma Chemical Co., St. Louis, Missouri) at pH 9.0 or pH 7.0 (see below), prior to incubation in the reaction mixture. The standard reaction mixture is similar to Novikoff and Goldfischer's (1968) modification of Graham and Karnovsky's (1966) original DAB procedure. The DAB reaction mixture was freshly prepared immediately before use, and contained: 5.0 ml of 0.05 *M* AMP buffer; 0.1 ml of 3% hydrogen peroxide (diluted from 30% superoxol); and 10 mg DAB (3,3′-diaminobenzidine tetrachloride; Sigma Chemical Co., St. Louis, Missouri). The final pH and temperature of the reaction mixture was adjusted to either pH 9.0 and 37°C, or pH 7.0 and 25°C, and incubation was allowed to proceed for 60 min. After incubation, the cells were rinsed several times in 0.1 *M* cacodylate buffer, pH 7.4. Osmication, dehydration, and embedding in Epon 812 followed the procedures described above.

Conditions of incubation (temperature and pH) were varied to favor the reactivity of specific enzymes with the DAB reaction mixture. When the final pH is adjusted to 9.0 and incubation occurs at 37°C, the DAB product of catalase activity can best be visualized; when the final pH is adjusted to 7.0 and incubation is at 25°C, the reactivity of cytochrome oxidase is favored (Novikoff and Goldfischer, 1969; Gerhardt and Berger, 1971). The following procedures were used with inhibitors of the DAB reaction product:

(a) Prior to incubation in the DAB reaction mixture, aliquots of cells fixed in glutaraldehyde and washed in AMP buffer were preincubated for 20 min in AMP buffer containing 0.01 *M* KCN at pH 7.0 and 25°C or pH 9.0 and 37°C. This treatment is followed by incubation in the DAB mixture containing 0.01 *M* KCN at either pH 7.0 and 25°C, or pH 9.0 and 37°C. KCN, under these conditions, inhibits both the reactivity of catalase and cytochrome oxidase with DAB.

(b) Prior to incubation in the DAB reaction mixture, aliquots of cells fixed in glutaraldehyde and washed in AMP buffer were preincubated for 20 min in AMP buffer containing 0.02 *M* aminotriazole (3-amino-1,2,4-triazole; Aldrich Chemical Co., Milwaukee, Wisconsin) (AT) at pH 7.0 and 25°C or pH 9.0 and 37°C. The cells were then incubated in the DAB mixture, which contained 0.02 *M* AT, and incubated at either pH 7.0 and 25°C or pH 9.0 and 37°C for 60 min. Under the latter conditions, AT specifically inhibits catalase activity (Vigil, 1970).

In all experiments DAB reactivity was monitored cytochemically by coprocessing germinating castor bean (*Ricinus communis* var. Baker 296) endosperm (a tissue known to exhibit high catalase activity and intense DAB staining), using the above procedures.

RESULTS

Since preliminary experiments disclosed a paucity of microbodies, and photometrically undetectable levels of catalase in glucose-supplemented cells, we concentrated our efforts on acetate-supplemented *Euglena gracilis* strain Z. Data on these cells are given in four major groupings: dark-grown cells, greening cells, greening cells treated with DCMU, and cells continuously cultured in the light.

Dark-Grown Cells

Non DAB-Treated Control

Microbodies are observed in cells cultured on both acetate and glucose supplemented media. Acetate-grown cells usually contain 3–5 microbodies in any particular section, with as many as 10 observed in some cells of a section. In contrast, many sections must be observed with glucose-supplemented cells in order to detect a microbody. The favored locations for microbodies are the cell periphery and regions near the gullet. An example is shown in Fig. 1 of the single membrane-bound, irregularly shaped structures, 0.4–1.4 μm in diameter, which we consider to be microbodies. Based on random counts of 200 sections of cells, there are approximately 300 mitochondria to each microbody in glucose-supplemented cells, while the ratio is about 25:1 in acetate-supplemented, dark-grown cells.

Photometric assays with cell-free extracts reveal 0.68 unit of catalase per milligram of protein in aerated acetate-grown cells. Although catalase activity could not be detected in acetate-supplemented cells grown in the absence of CO_2, reaeration with air containing CO_2 for 24 hr brought the catalase level back to 0.50 unit per milligram of protein.

DAB Treated

pH 9.0, 37°C. Incubation of dark-grown cells in the standard DAB mixture results in pronounced deposition of electron dense material in microbodies (Figs. 2 and 3). The DAB stain is still evident in the microbodies when H_2O_2 is omitted from

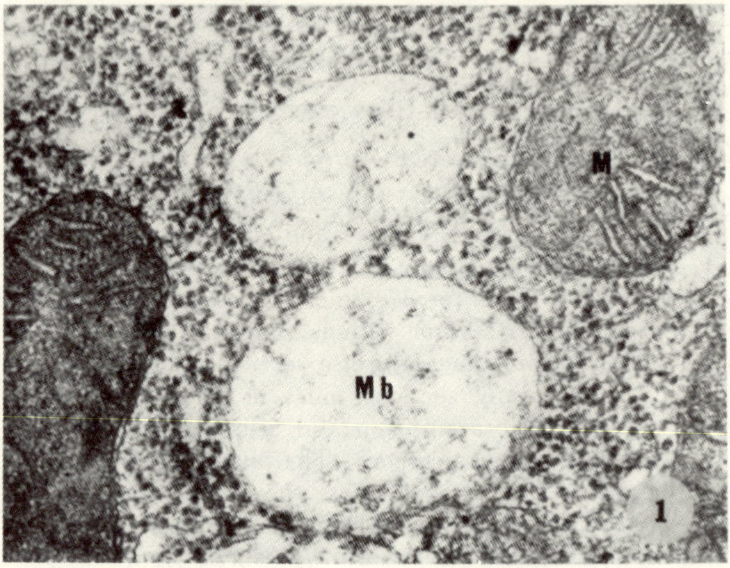

Fig. 1. Electron micrograph of non-DAB treated microbodies (M*b*) surrounded by mitochondria *(M)* in dark-grown, aerated, acetate-supplemented *Euglena gracilis* strain Z. x 50,000.

the DAB reaction mixture. These incubation conditions are optimal for catalase-mediated DAB reactivity. Deposition of the reaction product in microbodies is completely inhibited when 0.02 *M* AT or 0.01 *M* KCN is added to the reaction mixture.

pH 7.0, 25°C. Under this set of conditions neither the microbodies nor the mitochondria stain with DAB, and instead look like the non-DAB treated controls described above. That these conditions of incubation are optimal for detecting the presence of "conventional" cytochrome oxidase (see Discussion) was shown in the present work, in which intense DAB stain was deposited on the outer surface of the inner mitochondrial membrane of coprocessed endosperm of castor bean (*Ricinus communis*).

Greening Cells

Non DAB-Treated

Cells grown in either medium undergo an increase in number of microbodies; the increase begins after approximately 10 hr of greening, when the plastid has about 2 thylakoids. Based on random counts of 200 sections of acetate-supplemented cells, the number of microbodies has approximately doubled by 24 hr of greening (so that the ratio of mitochondria to microbodies is about 15:1), and after this time it remains fairly constant.

The application of Lück's (1963) assay to crude cell-free extracts of aerated, acetate-supplemented cells indicates that the concentration of catalase in 12-hr greening cells is 0.71 unit/mg protein, increases to 0.85 unit/mg protein in 24-hr greening cells, and further increases to 0.98 unit/mg protein in 48-hour greening cells. Catalase activity could not be detected cytochemically or photometrically in cells allowed to green in CO_2-free air. However, reaeration with air of dark-grown cells for 24 hr prior to greening, partially restores catalase activity to 0.46 unit/mg protein in 12-hr greening cells, 0.79 unit/mg protein in 24-hr greening cells, and 0.95 unit/mg protein in 48-hr greening cells.

DAB Treated

pH 9.0, 37°C. The mitochondria of cells that have undergone greening for 0–20 hr resemble those of the control or the dark-grown ones. However, between 20 and 24 hr of greening, a faint DAB reactivity is observed in some sections of mitochondria (Fig. 4). The DAB reaction product appears in the mitochondrial

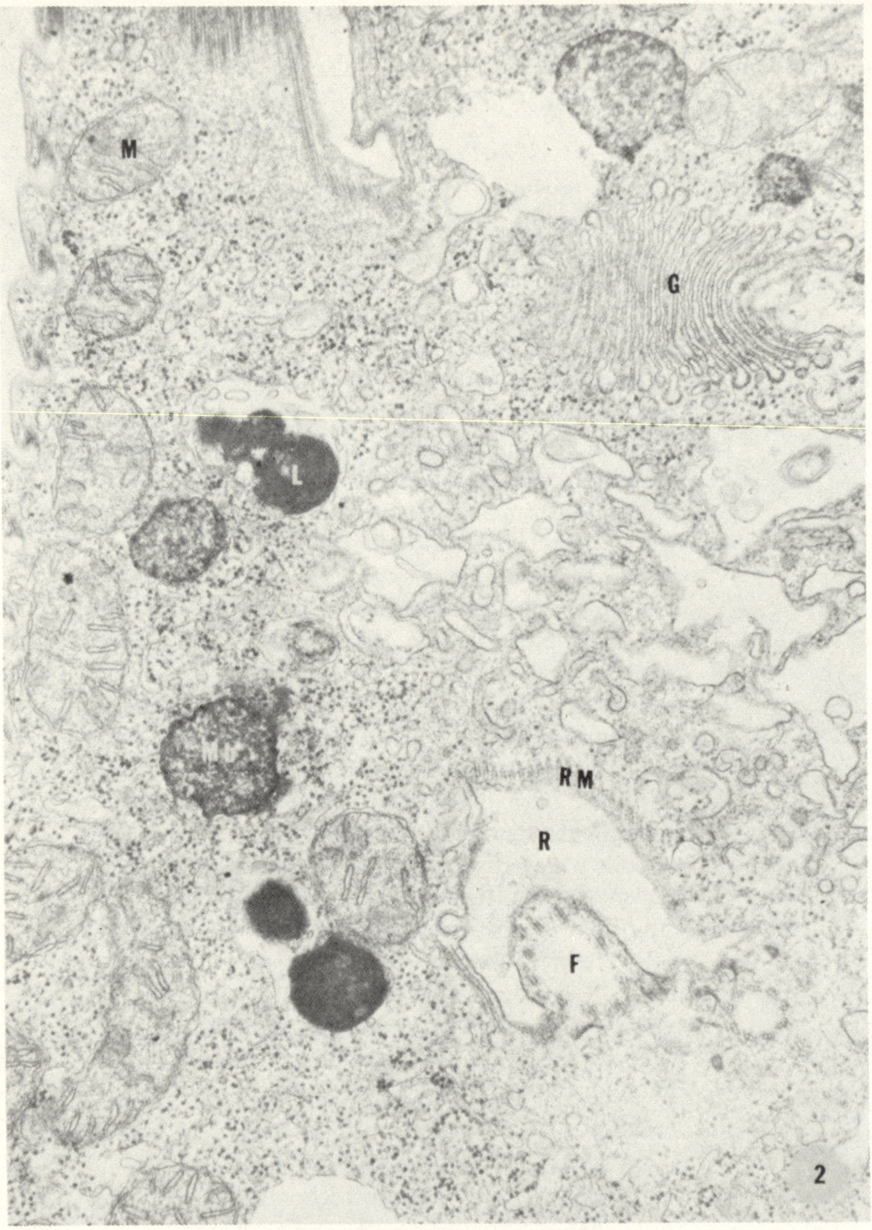

FIG. 2. Electron micrograph of DAB/H_2O_2 incubated (pH 9.0, 37°C), aerated, dark-grown, acetate-supplemented *Euglena gracilis* strain Z near gullet region. *F*, flagellum; *G*, Golgi body; *L*, lysosome; *M*, mitochondrion; *Mb*, microbody; *R*, reservoir; *RM*, reservoir microtubules. × 25,000.

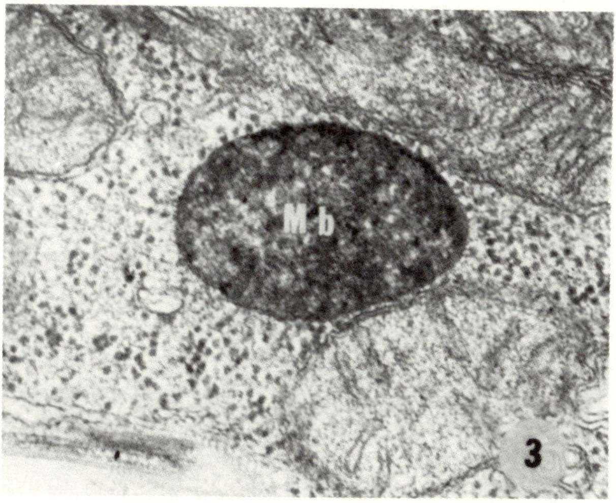

Fig. 3. DAB/H_2O_2 incubated (pH 9.0, 37°C) microbody *(Mb)* near periphery of dark grown, aerated, acetate-supplemented *Euglena gracilis* strain Z. x 50,000.

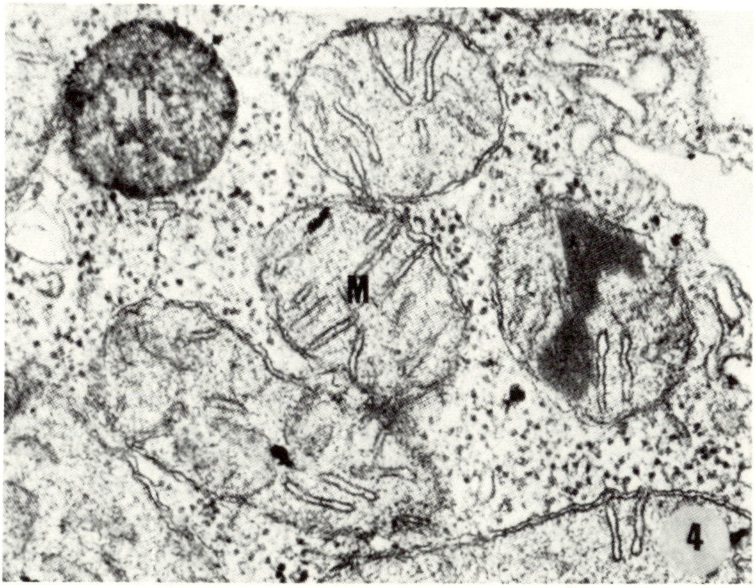

Fig. 4. Electron micrograph of DAB/H_2O_2 incubated (pH 9.0, 37°C), aerated, acetate-supplemented *Euglena gracilis* strain Z after 20 hrs. of greening, showing a microbody *(Mb)* and mitochondria *(M)*. x 40,000.

matrix, not in its membranes; also, it appears under incubation conditions that are not considered optimal for the visualization of cytochrome oxidase. At this stage of greening, the chloroplast is not fully developed, but is photosynthetically competent (Schiff, 1963; Brody *et al.*, 1965). However, sometime prior to 72 hr of greening, the intensity of the DAB stain decreases sharply. Electron-opaque material is not seen at all in the mitochondria of cells allowed to green for 72 (or 96) hr. The appearance of electron dense material in microbodies is again inhibited by preincubation in AT or KCN.

We attempted to better localize the DAB stain in mitochondria, using half the hydrogen peroxide and DAB concentrations of the standard DAB reaction mixture. However, the electron opaque appearance of the mitochondrial matrix was unchanged. Even in the absence of added hydrogen peroxide and in the presence of only half the standard amount of DAB, the stain appears unmodified.

KCN effectively inhibits the DAB reaction product in both microbodies and mitochondria. Aminotriazole, on the other hand, inhibits only the staining of microbodies; in the mitochondria of 24–48 hr greening cells, electron dense material is still apparent in the matrix (see Fig. 5). When dark-grown cells (grown in the presence or in the absence of CO_2) are allowed to green in the absence of CO_2, DAB reactivity fails to appear in the mitochondria.

pH 7.0, 25°C. At no time was there evidence of DAB deposition in mitochondria or microbodies of greening *Euglena* cells if the DAB reaction mixture was incubated at pH 7.0 and 25°C. In all such cases the mitochondria and microbodies appear as they do in the non-DAB treated cells. The data for DAB-treated greening cells are summarized in Table 1 (see Control columns).

DCMU-Treated Greening Cells

The results in this section are also summarized in Table 1.

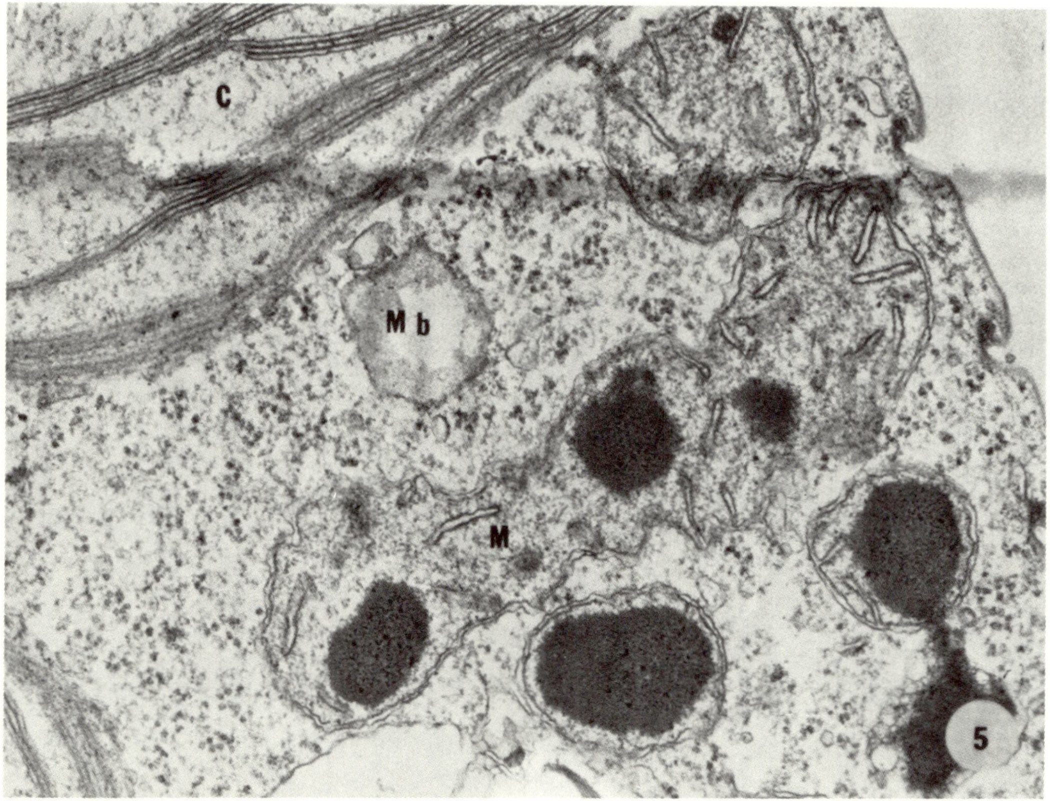

Fig. 5. Electron micrograph of DAB/H_2O_2 incubated (pH 9.0, 37°C) in aminotriazole, aerated, acetate-supplemented *Euglena gracilis* strain Z after 48 hr of greening, showing mitochondria (*M*), a microbody (*Mb*), and chloroplast (*C*). × 25,000.

pH 9.0, 37°C. Dark-grown acetate-supplemented cells were aerated with air and treated in the following manner:

(i) Six flasks were removed from the dark and DCMU added immediately to five flasks, before all six were placed in the light. Cells in each flask were allowed to green for increasing intervals of time, i.e., 18, 24, 40, 48, 72 hr. Aliquots from each flask were observed directly (control) or were treated with DAB, containing or not containing the inhibitors AT and KCN.

(ii) Five flasks of cells were removed from the dark and allowed to green for 18 hr, and then DCMU was added to 4 of the flasks; cells were allowed to green for an additional 6, 22, 30, or 54 hr (to yield total times in the light identical to cells in schedule i). Aliquots from each flask were observed directly or were treated with DAB, with and without the inhibitors AT or KCN.

(iii) Of four flasks of cells which had greened for 24 hr, DCMU was added to three; cells were allowed to green for additional 16, 24, or 48 hr to yield total times in the light of 40, 48, and 72 hr, respectively. The remaining procedures were as described above.

(iv) Of three flasks of cells which had greened for 40 hr, DCMU was added to two, and cells were allowed to green for an additional 8 or 32 hr, to yield a total of 48 or 72 hr in the light, respectively. The remaining procedures were as above.

pH 7.0, 25°C. The above schedules (i, ii, iii, iv) were repeated, but with cells incubated in the DAB reaction mixture at pH 7.0 and 25°C prior to fixation for electron microscopy.

From Table 1 (part i at pH 9.0, 37°C) it may be observed that the effect of DCMU treatment at zero time (i.e., be-

TABLE 1

RELATIVE INTENSITIES OF DAB STAIN IN MICROBODIES (Mb) AND MITOCHONDRIA (M) OF GREENING *Euglena* CELLS AS FUNCTIONS OF CONDITIONS OF INCUBATION AND TIME OF ADDITION OF DCMU[a]

Time of addition of DCMU	Hours in light	pH 9.0, 37° C						pH 7.0, 25° C					
		DAB		AT		CN		DAB		AT		CN	
		M	Mb	M	Mb	M	Mb	M	Mb	M	Mb	M	Mb
Not added, control	18	–	+	–	–	–	–	–	–	–	–	–	–
	24	+	+	+	–	–	–	–	–	–	–	–	–
	40	++	+	+	–	–	–	–	–	–	–	–	–
	48	+++	+	+	–	–	–	–	–	–	–	–	–
	72	–	+	–	–	–	–	–	–	–	–	–	–
(i) Zero time	18	–	+	–	–	–	–	–	–	–	–	–	–
	24	–	+	–	–	–	–	–	–	–	–	–	–
	40	–	+	–	–	–	–	–	–	–	–	–	–
	48	–	+	–	–	–	–	–	–	–	–	–	–
	72	–	+	–	–	–	–	–	–	–	–	–	–
(ii) After 18 hr of greening	24	–	+	–	–	–	–	–	–	–	–	–	–
	40	–	+	–	–	–	–	–	–	–	–	–	–
	48	–	+	–	–	–	–	–	–	–	–	–	–
	72	–	+	–	–	–	–	–	–	–	–	–	–
(iii) After 24 hr of greening	40	–	+	–	–	–	–	–	–	–	–	–	–
	48	–	+	–	–	–	–	–	–	–	–	–	–
	72	–	+	–	–	–	–	–	–	–	–	–	–
(iv) After 40 hr of greening	48	–	+	–	–	–	–	–	–	–	–	–	–
	72	–	+	–	–	–	–	–	–	–	–	–	–

[a] It is not intended to compare the relative intensities of DAB stain in mitochondria and microbodies. Intensities are relative only within the mitochondrial category; the + or – in the case of microbodies refers only to presence or absence.

fore cells are placed in the light) is to prevent the DAB stain from occurring in the mitochondrial matrix, at 24 hr or any later time of greening (compare with control). Thus, it would seem that the appearance of the DAB stain in the mitochondria is dependent upon an actively photosynthesizing chloroplast. Electron density in microbodies is unchanged in the presence of DCMU.

Although one can clearly see the DAB stain in control mitochondria allowed to green for 24 hr (see Control in Table 1), it is evident that when DCMU is added at 18 hr of greening, and cells allowed to green further, no DAB reactivity appears (see Table 1, part ii). DCMU was also added to 24-hr (Table 1, part iii) or 40-hr (Table 1, part iv) greening cells—the mitochondria of which exhibit intense DAB stain; when these cells were fixed 2 hr later, and subsequently examined, the DAB stain was no longer apparent.

Cells Cultured Continuously in the Light

Although the DAB stain is taken up by the microbodies of such cells at pH 9.0 and 37°C, it never appears in mitochondria. When the DAB reaction mixture is adjusted to pH 7.0 and the incubation temperature to 25°C, DAB reactivity is absent in both microbodies and mitochondria.

Application of Lück's (1963) photometric technique to cell-free extracts of acetate-supplemented cells aerated with air yields a catalase activity of 1.25 units/mg protein. Catalase activity could not be detected photometrically or cytochemically in acetate-supplemented cells continuously cultured in the light, in the absence of carbon dioxide. Reaeration with air for 24 hr partially restores catalase activity to 1.05 units/mg protein in 24 hr, and to 1.20 units/mg protein in 48 hr.

DISCUSSION

Hanzely *et al.* (1971) and Graves *et al.* (1971a) reported on the basis of fine structure, the presence of microbodies in a streptomycin-bleached strain of *Euglena gracilis* var. *bacillaris*. These cells were grown under gentle aeration of the inorganic medium of Cramer and Myers (1952) supplemented with acetate, ethanol, or glucose. Ethanol-grown cells were examined for catalase activity, both by Lück's (1963) photometric assay (on crude homogenates or particulate fractions), and by the cytochemical procedures of Frederick and Newcomb (1969) which utilizes DAB/H_2O_2. The results were negative with both assays. We were not working with the streptomycin-bleached variety *bacillaris* (SM-L1) which Graves and co-workers (1971a) used. However, as seen in the present work, under otherwise similar conditions of growth (i.e., in the presence of 2-carbon substrates and CO_2), catalase activity is detected in non-streptomycin-bleached strain Z. We offer the suggestion, in agreement with Levedahl (1968), that differences in metabolism probably result from permanent bleaching with streptomycin.

Lord and Merrett (1971), presumably on the basis of enzyme assays with appropriate fractions of sucrose graded cell extracts (results not shown) were also unable to detect catalase; their work was with *Euglena gracilis* strain Z, grown either phototrophically in air, on the inorganic medium of Cramer and Myers (1952), or grown heterotrophically in the dark (gas phase unknown) on the same medium supplemented with glucose.

Since at least one enzyme of the glyoxylate bypass, isocitrate lyase, is undetectable in various strains of *Euglena*, including Z, grown photoautotrophically (Haigh and Beevers, 1964), it may well be that such organisms have few microbodies or that their microbodies have undetectable levels of catalase. We have observed (as had Graves *et al.*, 1971a) that very few microbodies develop in cells grown in inorganic medium supplemented with glucose; although in the present work we have shown catalase to be present in such microbodies, their paucity makes detection by the photometric method unlikely.

Gerhardt and Berger (1971) working with two acetate flagellates (gas phase unknown)—one colorless (*Polytomella caeca*) and one green (*Chlorogonium elongatum*)—also looked for catalase activity in microbodies. Although cytochemically their DAB/H_2O_2 results were negative, positive results were achieved photometrically with fractions containing structures similar in morphology to microbodies, obtained from a linear sucrose gradient of crude particulate preparations. Perhaps their conditions of DAB incubation (pH and temperature) were not optimal, or possibly their half-hour period of incubation was too short to produce a catalase-mediated DAB stain.

Müller *et al.* (1968) have reported a catalase level of 3.06 units/mg protein for homogenates of the colorless ciliate *Tetrahymena pyriformis*, grown in the presence of acetate, while Lui *et al.* (1968) have found 0.16 unit/mg protein for homogenates of the green flagellate *Ochromonas malhamensis*. As seen in the Results section above, catalase levels in cell-free extracts of acetate-supplemented *Euglena gracilis* strain Z range from 0.68 unit/mg protein in dark-grown cells, to 1.25 units/mg protein in cells cultured continuously in the light.

In the present work we note that acetate-grown cells have no detectable catalase when grown in CO_2-free air, but do, when grown in the presence of CO_2. Furthermore, addition of CO_2 to the gas phase results in restoration of catalase activity. Therefore, we speculate that CO_2 exerts a regulatory influence on the activity (or possibly, the presence) of this enzyme system.

It is interesting to note that while in higher plants, the oxidation of glycolic acid is catalyzed by glycolic acid oxidase—which transfers electrons to oxygen—it has been reported that in algae, the glycolate-oxidizing enzyme is a dehydrogenase and does not produce hydrogen peroxide (Codd *et al.*, 1969; Nelson and Tolbert, 1970). Lord and Merrett (1971) reported an increase in the specific activity of this dehydrogenase upon alteration of the gas phase from 5.0% to 0.03% CO_2 in phototrophically grown *Euglena gracilis* strain Z. The particulate nature of this enzyme was demonstrated by Lord and Merrett (1971), and by Graves *et al.* (1971b); Graves, furthermore reports its location in microbodies (personal communication). In view of our findings that catalase-mediated oxidative polymerization of DAB occurs—with or without exogeneous H_2O_2—it would be worthwhile to look for a "higher plant-like enzyme" (capable of producing H_2O_2) in microbodies of aerated, acetate-grown cells.

The presence of glyoxylate bypass enzymes in dark-grown acetate-supplemented cells of *Euglena gracilis* strain Z (Haigh and Beevers, 1962; Reeves *et al.*, 1962), coupled with the cytological finding that there are more microbodies in such cells than those grown on glucose (see also Graves *et al.*, 1971a), makes it likely that most (if not all) of the microbodies found in the former are glyoxysomes. The question arises whether the larger population of microbodies in greening cells represent an increase in glyoxysomes or the development of an enzymatically different type of microbody. Since the dark fixation of $^{14}CO_2$ by green var. *bacillaris*, growing in the presence of acetate, differs from that of the streptomycin bleached form (SM-L1), in that the former not only strongly labels succinate but also glycine and serine (see Levedahl, 1968), there exists the possibility that such green cells have a "leaflike" peroxisomal type of glycolic acid metabolism. Enzyme studies, now under way, will have to be completed before the exact nature of the incremented fraction of microbodies can be ascertained.

Palmer and Togasaki (1971) have studied ^{14}C-acetate metabolism in *Pandorina* and note that its light-stimulated uptake is diminished by aeration with CO_2-free air. They have also shown that lipid synthesis is dependent upon both light and photosynthesis—in the dark, or in the presence of DCMU, there is severe inhibition of ^{14}C incorporation into lipid.

Gerhardt and Berger (1971) observed that the mitochondrial cristae of their acetate flagellates stain with DAB, especially at neutral pH, and attribute this reactivity to cytochrome oxidase or peroxidase. We have not be able to detect a similar pattern of deposition in *Euglena* in the present study, although we have seen it in the outer surface of the inner mitochondrial membranes of coprocessed *Ricinus* endosperm. In this regard, we note that although Perini *et al.* (1964a, b) found an "a" type cytochrome (605) in dark-grown *Euglena*, they concluded that their evidence was insufficient to characterize it as "a_3". Krawiec and Eisenstadt (1970) and Lord and Merrett (1971) reported being unable to locate *Euglena* mitochondria by cytochrome oxidase assays. However, Sharpless and Butow (1970), working with a permanently bleached strain of *Euglena gracilis*, estimated from low temperature (77°K) difference spectra that an average concentration of cytochrome oxidase of 0.25 nmoles/mg of mitochondrial protein was present.

Our observation that the time dependence of the mitochondrial DAB stain (its visualization at 20 hr of greening, its maximum intensity at about 48 hr of greening, and its disappearance by 72 hr of greening) approximates the first derivative of the greening curve, seems to closely relate it either to a light requirment per se or to the rate of synthesis of plastid constituents. Schiff *et al.* (1967) observed that dark-grown cells can be light induced to form chloroplasts in the presence of the photosynthesis inhibitor DCMU, and suggested that the developing plastid utilizes energy, not from photosynthesis but from the cytoplasm (see also Schiff, 1971). Since the appearance of detectable levels of the mitochondrial enzyme(s) is inhibited by DCMU, and since its activity is deficient in CO_2-free air, its functioning may be photosynthesis-dependent. Inasmuch as, in the absence of an added electron donor, DCMU makes both photosystems I and II inoperative, further experiments will have to be performed to determine whether the enzyme activity is a product of one or both of these systems.

The transitory nature of the DAB mitochondrial stain seems to connect it with the conversion of a substrate produced in the dark and/or the synthesis of a compound needed for new metabolism in the light. In this regard, one should note that Smillie *et al.* (1963) have observed a transient increased activity of several enzymes of glucan metabolism and respiration, when dark-adapted *Euglena gracilis* cells are exposed to light. It is also possible that the activity of the mitochondrial enzyme(s) reported in the present work is associated with a modified lipid metabolism. Erwin and Bloch (1962) have shown that *Euglena gracilis* grown in the light on organic medium has large amounts of α-linolenic acid (mainly in the glycolipids of the chloroplasts), while *Euglena* grown in the dark has C-20, C-22, and C-24 polyenoic fatty acids of the γ-linolenic type (mainly in the phospholipid portion of their membranes). The nature of the enzyme which results in the mitochondrial DAB reaction is under investigation.

Note added in proof. Graves (personal communication) has reported that when he uses a modification of our method, i.e., bubbles 1 % CO_2 through streptomycin-bleached *Euglena* (SM-L1), he is able to detect small amounts of catalase in cell homogenates.

We wish to thank Dr. K. Lyser for her gracious cooperation in the electron microscope studies.

REFERENCES

BREIDENBACH, R. W., and BEEVERS, H. (1967). Association of the glyoxylate cycle enzymes in a novel subcellular particle from castor bean endosperm. *Biochem. Biophys. Res. Commun.* **27,** 462–469.

BREIDENBACH, R. W., KAHN, A., and BEEVERS, H. (1968). Characterization of glyoxysomes from castor bean endosperm. *Plant Physiol.* **43,** 705–713.

BRODY, M., BRODY, S. S., and LEVINE, J. H. (1965). Fluorescence changes during chlorophyll formation in *Euglena gracilis* (and other organisms) and an estimate of lamellar area as a function of age. *J. Protozool.* **12,** 465–475.

CODD, G. A., LORD, J. M., and MERRETT, M. J. (1969). The glycollate oxidising enzyme of Algae. *FEBS Lett.* **5,** 341–342.

COGGESHALL, R. E. (1965). A fine structure analysis of the ventral nerve cord and associated sheath of

Lumbricus terrestris. J. Comp. Neurol. **125,** 393-438.

Cramer, M., and Myers, J. (1952). Growth and photosynthetic characteristics of *Euglena gracilis. Arch. Mikrobiol.* **17,** 384-402.

de Duve, C. (1965). Functions of microbodies (peroxisomes). *J. Cell Biol.* **27,** 25A-26A.

de Duve, C. and Baudhuin, P. (1966). Peroxisomes (microbodies and related particles). *Physiol. Rev.* **46,** 323-357.

Erwin, J., and Bloch, K. (1962). The α-linolenic content of some photosynthetic microorganisms. *Biochem. Biophys. Res. Commun.* **9,** 103-108.

Frederick, S. E., and Newcomb, E. H. (1969). Cytochemical localization of catalase in leaf microbodies (peroxisomes). *J. Cell Biol.* **43,** 343-353.

Frederick, S. E., Newcomb, E. H., Vigil, E. L., and Wergin, W. P. (1968). Fine structural characterization of plant microbodies. *Planta* **81,** 229-252.

Gänsler, H., and Rouiller, C. (1956). Modifications physiologiques et pathologiques du chrondriome. *Schweiz. Z. Allgem. Pathol. Bacteriol.* **19,** 217-243.

Gergis, M. S. (1971). The presence of microbodies in three strains of *Chlorella. Planta* **101,** 180-184.

Gerhardt, B., and Berger, C. (1971). Microbodies und Diaminobenzidin-Reaktion in den Acetat-Flagellaten *Polytomella caeca* und *Chlorogonium elongatum. Planta* **100,** 155-166.

Graves, L. B., Jr., Hanzely, L., and Trelease, R. N. (1971a). The occurrence and fine structural characterization of microbodies in *Euglena gracilis. Protoplasma* **72,** 141-152.

Graves, L. B., Trelease, R. N., and Becker, W. M. (1971b). Particulate nature of glycolate dehydrogenase in *Euglena*: Possible localization in microbodies. *Biochem. Biophys. Res. Commun.* **44,** 280-286.

Graham, R. C., Jr., and Karnovsky, M. J. (1966). The early stages of injected horseradish peroxidase in the proximal tubules of mouse kidney: Ultrastructural cytochemistry by a new technique. *J. Histochem. Cytochem.* **14,** 291-302.

Haigh, W. G., and Beevers, H. (1962). Acetate metabolism in algae. *Plant Physiol.* **37,** Suppl., **lx**.

Haigh, W. G., and Beevers, H. (1964). The occurrence and assay of isocitrate lyase in algae. *Arch. Biochem. Biophys.* **107,** 147-151.

Hanzely, L., Graves, L. B., Jr., and Molsen, D. (1971). Development of a multilaminate core in microbodies of *Euglena gracilis* var. *bacillaris* (SM-L1). Reprinted from: "29th Ann. Proc. Electron

Microscopy Soc. Amer." (C. J. Arceneaux, ed.), pp. 57–59. Boston, Massachusetts.

Hogg, J. F. (1969). Peroxisomes in *Tetrahymena* and their relation to gluconeogenesis. *Ann. N.Y. Acad Sci.* **168,** 281–291.

Kisaki, T., and Tolbert, N. E. (1969). Glycolate and glyoxylate metabolism by isolated peroxisomes or chloroplasts. *Plant Physiol.* **44,** 242–250.

Krawiec, S., and Eisenstadt, J. M. (1970). Ribonucleic acids from the mitochondria of bleached *Euglena gracilis.* I. Isolation of mitochondria and extraction of nucleic acid. *Biochim. Biophys. Acta* **217,** 120–131.

Levedahl, B. H. (1968). Heterotrophic CO_2 fixation in *Euglena. In* "The Biology of Euglena" (D. E. Buetow, ed), Vol. II, pp. 85–96. Academic Press, New York.

Lord, J. M. and Merrett, M. J. (1971). The intracellular localization of glycollate oxidoreductase in *Euglena gracilis. Biochem. J.* **124,** 275–281.

Lowry, O. H., Rosebrough, N. J., Farr, A. L., and Randall, R. J. (1951). Protein measurement with Folin phenol reagent. *J. Biol. Chem.* **193,** 265–275.

Lück, H. (1963). Catalase. *In* "Methods of Enzymatic Analysis" (H. U. Bergmeyer, ed.), pp. 885–888. Academic Press, New York.

Lui, N. S. T., Roels, O. A., Trout, M. E., and Anderson, O. R. (1968). Subcellular distribution of enzymes in *Ochromonas malhamensis. J. Protozool.* **15,** 536–542.

Mollenhauer, H. H., Morré, D. J., and Kelley, A. G. (1966). The widespread occurrence of plant cytosomes resembling animal microbodies. *Protoplasma* **62,** 44–52.

Müller, M. (1969). Peroxisomes of protozoa. *Ann. N. Y. Acad. Sci.* **168,** 292–301.

Müller, M., Hogg, J. F., and de Duve, C. (1968). Distribution of tricarboxylic acid cycle enzymes and of glyoxylate cycle enzymes between mitochondria and peroxisomes in *Tetrahymena pyriformis. J. Biol. Chem.* **243,** 5385–5395.

Nelson, E. B., and Tolbert, N. E. (1970). Glycolate dehydrogenase in green algae. *Arch. Biochem. Biophys.* **141,** 102–110.

Novikoff, A. B., and Goldfischer, S. (1968). Visualization of microbodies for light and electron microscopy. *J. Histochem. Cytochem.* **16,** 507.

Novikoff, A. B., and Goldfischer, S. (1969). Visualization of peroxisomes (microbodies) and mitochondria with diaminobenzidine. *J. Histochem. Cytochem.* **17,** 675–680.

PALMER, E. G., and TOGASAKI, R. K. (1971). Acetate metabolism by an obligative phototrophic strain of *Pandorina morum*. *J. Protzool.* **18,** 640–644.

PERINI, F., KAMEN, M. D., and SCHIFF, J. A. (1964a). Iron-containing proteins in *Euglena*. I. Detection and characterization. *Biochim. Biophys. Acta* **88,** 74–90.

PERINI, F., SCHIFF, J. A., and KAMEN, M. D. (1964b). Iron-containing proteins in *Euglena*. II. Functional localization. *Biochim. Biophys. Acta* **88,** 91–98.

REEVES, H. C., KADIS, S., and AJL, S. (1962). Enzymes of the glyoxylate by-pass in *Euglena gracilis*. *Biochim. Biophys. Acta* **57,** 403–404.

RHODIN, J. (1954). Correlation of ultrastructural organization and function in normal and experimentally changed proximal convoluted tubules cells of the mouse kidney. Aktiebolaget Godvil, Stockholm.

SCHIFF, J. A. (1963). Oxygen exchange by *Euglena* cells undergoing chloroplast development. *Carnegie Inst. Wash. Yearb.* 375–378.

SCHIFF, J. A. (1971). The informational and nutritional requirements of cellular organelles. *Genetics, Stadler Symp.* **3,** 89–113.

SCHIFF, J. A., ZELDIN, M. H., and RUBMAN, J. (1967). Chlorophyll formation and photosynthetic competence in *Euglena* during light-induced chloroplast development in the presence of 3,(3,4-dichlorophenyl)1,1-dimethylurea (DCMY). *Plant Physiol.* **42,** 1716–1725.

SHARPLESS, T. K., and BUTOW, R. A. (1970). Phosphorylation sites, cytochrome complement, and alternate pathways of coupled electron transport in *Euglena gracilis* mitochondria. *J. Biol. Chem.* **245,** 50–57.

SMILLIE, R. M., EVANS, W. R., and LYMAN, H. (1963). Metabolic events during the formation of a photosynthetic from a non-photosynthetic cell. *Brookhaven Symp. Biol.* **16,** 89–108.

TOLBERT, N. E., and YAMAZAKI, R. K. (1969). Leaf peroxisomes and their relation to photorespiration and photosynthesis. *Ann. N. Y. Acad. Sci.* **168,** 325–341.

TOLBERT, N. E., OESER, A., KISAKI, T., HAGEMAN, R. H., and YAMAZAKI, R. K. (1968). Peroxisomes from spinach leaves containing enzymes related to glycolate metabolism. *J. Biol. Chem.* **243,** 5179–5184.

VIGIL, E. L. (1970). Cytochemical and developmental changes in microbodies (glyoxysomes) and related organelles of castor bean endosperm. *J. Cell Biol.* **46,** 435–454.

Watson, M. L. (1958). Staining of tissue sections for electron microscopy with heavy metals. *J. Biophys. Biochem. Cytol.* **4,** 475–478.

pH CONTROL OF THE CHLOROPHYLL *a* FLUORESCENCE IN ALGAE

GEORGE PAPAGEORGIOU AND GOVINDJEE

SUMMARY

The pH of the suspension medium was found to have a remarkable influence on the "slow" (min) time course of Chlorophyll *a* fluorescence yield in the green alga *Chlorella pyrenoidosa* and in the blue-green alga *Anacystis nidulans*. In Chlorella, the decay of fluorescence yield, in the 1- to 5-min region, is strongly retarded at alkaline pH; this decay rate shows an optimum at pH 6–7. In Anacystis, the rise of fluorescence yield, in the same time range, is decreased optimally at pH 6–7; poisoning with 3(3,4-dichlorophenyl)-1,1-dimethylurea reverses the direction of this pH effect. These observations suggest a correlation of the H^+ status (or the processes associated with it such as photophosphorylation and resulting conformational changes) of the chloroplast to the yield of chlorophyll *a* fluorescence *in vivo*.

Light-dependent proton uptake by cloroplasts is related to the overall process of photosynthesis[1,2]. Similar movements of protons and of other cations have also been demonstrated in suspensions of whole algal cells[3–7]. In the latter case, the observed pH change is the result both of a light-induced proton pump, and of the migration of undissociated carbonic acid from the medium to the CO_2-depleted photosynthetic sites of the interior. Electron transport inhibitors, and uncouplers of photophosphorylation that affect the normal course of photosynthesis suppress the light-induced ion fluxes.

The ionic composition of the algal cell interior exerts a 2-fold influence on the conformation of the chloroplasts and thylakoids. It may bring about the osmotic shrinkage or distention of these entities, or it may modify the fine structure of the lamellae by regulating the charge of its ionizable groups (*e.g.* phospholipids, polar amino acid residues). In either case it affects various optical properties of the cell, and especially the fluorescence of chlorophyll *a*. Changes in the fluorescence spectrum, the fluorescence yield, and the kinetics of the fluorescence yield on the addition of electrolytes to the medium have been observed by several investigators. In Euglena chloroplast fragments, BRODY *et al.*[8] demonstrated a suppression of the far-red (715–736 nm) chlorophyll *a* fluorescence band, at low temperature, in the presence of concentrated salts, or at pH values away from neutrality. MURATA[9] and MURATA *et al.*[10] have shown that this suppression can be caused by dilute Mg^{2+} salts, and that,

Abbreviation: DCMU, 3(3,4-dichlorophenyl)-1,1-dimethylurea.

at 77°K, it is accompanied by a relative increase of chlorophyll *a* fluorescence bands at 684 nm and at 695 nm with respect to that at 715–736 nm. In chloroplasts of *Phytolacca americana*, HOMANN[11] observed an increase of the P level of the fast fluorescence transient on the addition of dilute Mg^{2+}; this increase was also present in preparations treated with 3(3,4-dichlorophenyl)-1,1-dimethylurea (DCMU) and sodium dithionite. On the other hand, P. MOHANTY (personal communication) and DE KOUCHKOVSKY[12] showed independently that concentrated salts suppress the amplitude of the fast fluorescence transient (*i.e.* the P level) in suspensions of whole Chlorella cells, and influence in general, the kinetic course of the chlorophyll *a* fluorescence yield. In this communication we report the effect of the proton concentration of the suspension medium on the slow kinetics of chlorophyll *a* fluorescence in *Chlorella pyrenoidosa* (unicellular chlorophyte) and *Anacystis nidulans* (unicellular cyanophyte).

The algae were grown in inorganic medium, in continuous illumination with white light of low intensity, as described elsewhere[13]. 3- to 6-day-old cultures were transferred into a buffered solution (0.02 M Tris–HCl, 0.04 M NaCl, or 0.01 M phosphate buffer), and the absorbance of the suspension at the red Chl *a* absorption maximum was adjusted to 0.5 for 1-cm light path. The total chlorophyll content of the Chlorella samples was 18 μg/ml, while that of the Anacystis samples was 8μg/ml. These values were determined with the procedure of ARNON[14]. The pH of the samples was adjusted by adding HCl or NaOH to the suspension. In spite of the poor buffering capacity of Tris below pH 6.5, the pH of the samples, assayed before and after the fluorescence measurements, did not change. The instrument and the technique used in recording the slow change of the chlorophyll *a* fluorescence yield in these algae has been detailed elsewhere[15–17]. The reproducibility of the kinetic pattern was ensured by a 15-min adaptation of the cells to darkness that preceded all measurements. The fluorescence time course data are presented in terms of the ratio $f_t = F_t/F_3''$ (relative fluorescence yield), where F_t and F_3'' denote the fluorescence intensity at t sec and

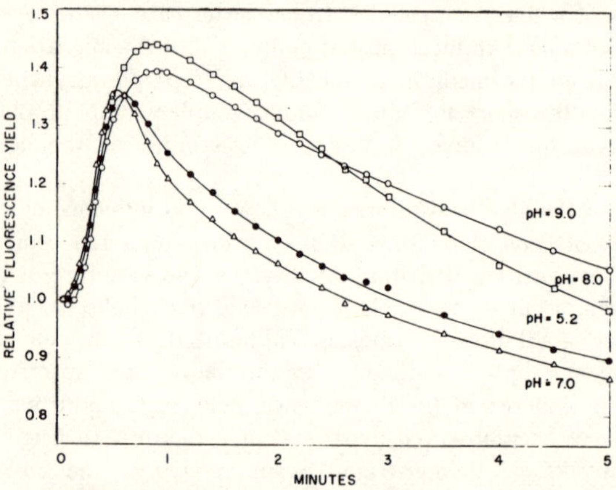

Fig. 1. Time course of the relative fluorescence yield ($f_t = F_t/F_3''$) of chlorophyll *a* in *Chlorella pyrenoidosa* at different pH values of the suspension medium. The fluorescence kinetics are recorded with continuous exciting illumination; 15-min dark adaptation preceded these measurements. Fluorescence excitation: λ = 480 nm; half-band width, 6.6 nm; incident intensity, 650 ergs·cm^{-2}·sec^{-1}. Fluorescence observation: λ = 685 nm; half-band width 16.5 nm; Corning sharp cut off filter, C.S. 2-62.

at 3 sec, respectively. F_3'' corresponds to the "S level" at the end of the fast fluorescence transient. Further details are given in the legends to the figures.

Fig. 1 shows the time course of the relative fluorescence yield of chlorophyll *a* in Chlorella at various proton concentrations of the suspending phase. The rise portion of the depicted fluorescence transient appears to be insensitive to the proton concentration of the medium, while the decay portion is markedly retarded in the alkaline region. The greater amplitude of the transient and the displacement of the maximum toward longer times at high pH values can be traced to the different sensitivities of the fluorescence rise and decay processes to the pH of the medium. A decelerated decay allows the forward changes to proceed further resulting in transients of greater amplitudes and with maxima displaced to longer times. These results suggest that different physico-chemical causes underly the rise and the decay portions of the slow fluorescence change in Chlorella.

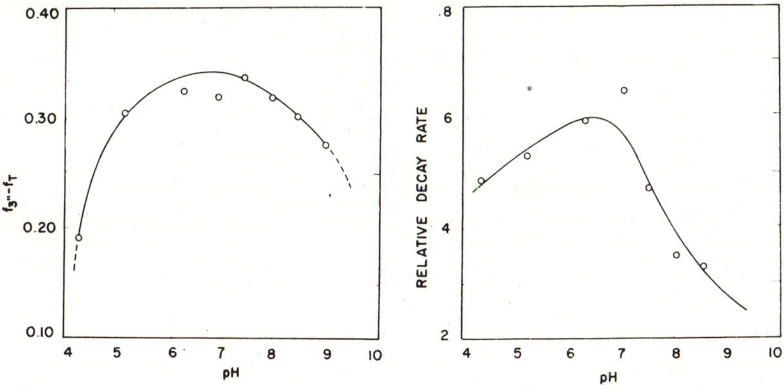

Fig. 2. The fluorescence change $f_3''-f_T$ (left) and the decay rate of the fluorescence time course (right) of chlorophyll *a* in *Chlorella pyrenoidosa* as a function of the pH of the suspension medium; details as in Fig. 1.

The difference $f_3''-f_T$ (Fig. 2, left) and the rate of fluorescence decay after the induction maximum (Fig. 2, right) exhibit optima in the pH range of 6–7. f_T is lower than f_3'', and denotes the final level of the relative fluorescence yield at the end of the induction process. The inverse of the time interval at which f_t crosses the level 1.0 (the level equal to that at S) of the ordinate is taken as a measure of the fluorescence decay rate. This is justifiable as the rise portion of the transient appears to be independent of the pH (Fig. 1).

In the blue-green alga *Anacystis nidulans*, the chlorophyll *a* fluorescence yield rises from the level S to a higher plateau (M), and then decays very slowly to final level T (ref. 18). In Anacystis, in contrast to Chlorella, the oxygen evolution inhibitor DCMU does not prevent the slow fluorescence change, but instead, it enhances it. The phenomenology and the implications of these effects have been discussed elsewhere[16]. In the present study, we found the fluorescence yields at the "S" levels of both normal and DCMU-poisoned Anacystis to be pH insensitive, and the observed pH effects were limited to the light induced increments (S to M rise) only. (See Fig. 3 for DCMU-poisoned Anacystis.) The pH curves of the maximum amplitude f_M-f_3''

(f_M is the plateau level) are given in Fig. 4 for a normal (left) and a DCMU-poisoned sample (right). The fluorescence induction amplitude of normal Anacystis has a minimum at pH 6–7, while the DCMU-poisoned cells exhibit a maximum in the same pH range.

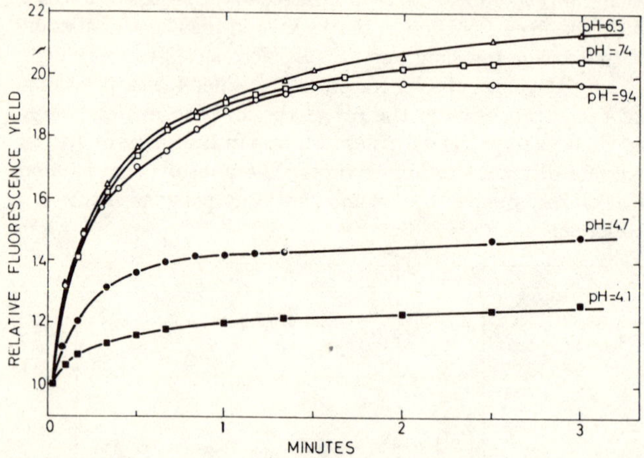

Fig. 3. Time course of the relative fluorescence yield ($f_t = F_t/F_3''$) of chlorophyll *a* in DCMU-poisoned *Anacystis nidulans* at different pH values of the suspension medium. The fluorescence kinetics are recorded with continuous exciting illumination; 15-min dark adaption preceded these measurements. DCMU, $5 \cdot 10^{-5}$ M. Fluorescence excitation; λ = 590 nm; half-band width, 16.5 nm; incident intensity, $4.1 \cdot 10^3$ ergs.cm^{-2}·sec^{-1}. Fluorescence observation : λ = 685 nm; half-band width, 6.6 nm; Corning sharp cut off filter, C.S. 2-58.

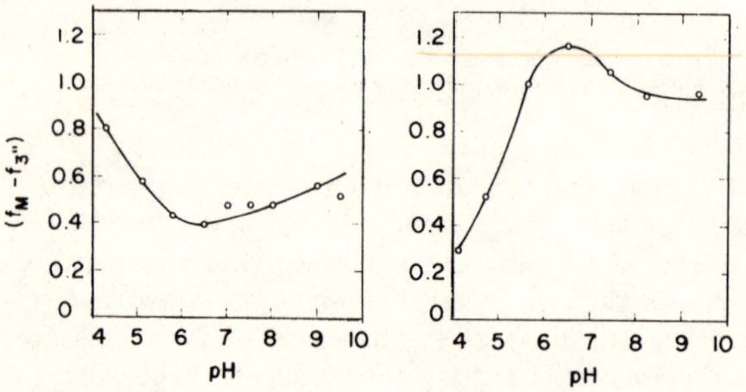

Fig. 4. The fluorescence change $f_M - f_3''$ of chlorophyll *a* in normal *Anacystis nidulans* as a function of the pH of the suspension medium (left); the same in DCMU-poisoned Anacystis (right); details as in Fig. 3.

These results underscore a direct or an indirect importance of the proton concentration of the medium for the slow kinetics of chlorophyll *a* fluorescence in whole cells of unicellular algae. Although we do not know the pH at the photosynthetic sites in the interior of the cell, we consider that the proton concentration of the suspension medium exerts an effect both by the passive diffusion of undissociated carbonic acid and by the operation of a light-dependent proton pump[6,7].

In the algae studied here, as well as in isolated higher plant chloroplasts[19, 20], phenomena associated with light-induced shifts of ion concentrations have optima in the physiological pH range of 6–7. The results presented here illustrate the sensitivity of the fluorescence induction to pH (see review on fluorescence induction[21]). The limitation of the pH effect in Chlorella to the decay portion of the slow fluorescence change resembles similar limitations in the action of carbonyl cyanide *p*-trifluoromethoxy phenylhydrazone[17] and KCN[22]. Evidently, the physical causes that lead to the rise and the decay of chlorophyll *a* fluorescence in Chlorella are distinguishable. According to previous data[16,21], only the cyclic electron transport contributes to the slow fluorescence change of DCMU-poisoned Anacystis, while in normal samples both the non-cyclic and the cyclic transports are effective. This differentiation may be the basis of the opposite trends in the pH curves of the fluorescence induction amplitude in normal and poisoned Anacystis.

In conclusion, the above results show clearly that the pH of the suspension medium has a remarkable effect on the slow time course of chlorophyll *a* fluorescence yield in algae, and thus suggest a correlation of the H^+ status of the chloroplast to chlorophyll fluorescence *in vivo*. We do not yet know whether this correlation is direct or indirect *via* the associated processes such as phosphorylation and the resulting conformational changes (see ref. 22).

ACKNOWLEDGEMENT

We gratefully acknowledge support by the National Science Foundation (U.S.A.) and the Greek Atomic Energy Commission.

REFERENCES

1 A. T. Jagendorf and E. Uribe, *Brookhaven Symp. Biol.*, 19 (1967) 215.
2 L. Packer and A. R. Crofts, in D. R. Sanadi, *Current Topics in Bioenergetics*, Vol. 2, Academic Press, New York, 1967, p. 23.
3 J. T. Cummins, J. A. Strand, and B. E. Vaughan, *Biochim. Biophys. Acta*, 173 (1969) 198.
4 W. J. Vredenberg, in H. Metzner, *Progress in Photosynthesis Research*, Vol. 2, Tübingen, 1969, p. 923.
5 W. J. Vredenberg, *Biochem. Biophys. Res. Commun.* 37 (1969) 785.
6 A. Ben-Amotz and B. Z. Ginzburg, *Biochim. Biophys. Acta*, 183 (1969) 144.
7 S. Schuldiner and I. Ohad, *Biochim. Biophys. Acta*, 180 (1969) 165.
8 S. S. Brody, C. A. Ziegelmair, A. Samuels, and M. Brody, *Plant Physiol.*, 41 (1966) 1709.
9 N. Murata, *Biochim. Biophys. Acta*, 189 (1969) 171.
10 N. Murata, H. Tashiro, and A. Takamiya, *Biochim. Biophys. Acta*, 197 (1970) 250.
11 P. Homann, *Plant Physiol.*, 44 (1969) 932.
12 Y. De Kouchkovsky, *Abstr. 8th Intern. Congr. Biochem.* (1970) p. 139.
13 Govindjee and E. Rabinowitch, *Biophys. J.*, 1 (1960) 73.
14 D. I. Arnon, *Plant Physiol.*, 24 (1949) 1.
15 Govindjee, in J. B. Thomas and J. H. C. Goedheer, *Currents in Photosynthesis*, Ad Donker, Rotterdam, 1966, p. 93.
16 G. Papageorgiou and Govindjee, *Biophys. J.*, 8 (1968) 1299.
17 G. Papageorgiou and Govindjee, *Biophys. J.*, 8 (1968) 1316.
18 G. Papageorgiou and Govindjee, *Biophys. J.*, 7 (1967) 375.
19 J. Neumann and A. T. Jagendorf, *Arch. Biochem. Biophys.*, 107 (1964) 109.
20 L. Packer, D. W. Deamer, and A. R. Crofts, *Brookhaven Symp. Biol.*, 19 (1966) 281.
21 Govindjee and G. Papageorgiou, in A. C. Giese, *Photophysiology*, Vol. 6, Academic Press, New York, 1971, in the press.
22 E. C. Wassink, and E. Katz, *Enzymologia*, 6 (1939) 145.

QUANTIZATION OF A FLUX RATIO IN CHAROPHYTES?

G. P. FINDLAY*, A. B. HOPE* AND N. A. WALKER**

SUMMARY

1. The ratio of the Cl^- influx into the vacuole to the Cl^- influx into the whole cell has been measured for a large number of internodal cells of *Chara corallina*. The distribution of the values of this ratio shows no tendency to cluster about preferred values in integer ratios with each other.

2. The claim of MacRobbie[1], that values of this ratio for *Nitella translucens* do cluster in this way, is shown to be unsupported by her data.

3. It is concluded that there is no evidence for the phenomenon of "quantization"[1] of this flux ratio.

INTRODUCTION

Recently MacRobbie[1,2] has discussed the kinetics of the uptake of Cl^- by cells of Nitella and Tolypella. Ion uptake in plant cells has generally been described in terms of the three-compartment model[3,4] (in which outside, cytoplasm and vacuole are the compartments), though recently Larkum[5] and Larkum and Pring[6] have considered four-compartment models in which the chloroplasts form the fourth compartment. MacRobbie[1,2] has concluded from a number of features of the kinetics, particularly at short times of uptake, that both three- and four-compartment models of the cell could be rejected, and has argued instead that salt uptake can be described in terms of a model including pinocytosis or vesicular transport. Since this concept could be of fundamental importance in the study of ion uptake, we are examining some of the kinetic features she has described.

Here, we discuss the phenomenon of "quantization"[1]. When charophyte cells are allowed to take up radioactive Cl^- for a short time (about 10 min), the distribution of this Cl^- between cytoplasm and vacuole may be determined[1,2]. MacRobbie[1] contended that the proportion of Cl^- reaching the vacuole, (M_{ov}/M_T)*, tended to take on, in any single experiment, discrete values related in the ratios 1:2:3:4·····. Thus in one experiment on 22 cells the results for M_{ov}/M_T were interpreted as clustering round the following series of values of $n\alpha$: 0.051, 0.102, 0.153, ····, 0.306; while

Abbreviation: HEPES, N-2-hydroxyethylpiperazine N'-2'-ethane sulphonc acid.

* For ease of cross-reference we will use the symbols of MacRobbie[1] throughout this paper, i e. M_{ov}, influx to vauole from outside solution; M_T, influx into cell from outside solution; α, a constant determined for each experiment; n, an integer constant assigned to a group of values of M_{ov}/M_T. The constants α and n occur in the equation of MacRobbie $M_{ov} = n\alpha M_T$.

in another experiment on 18 similar cells they were said to cluster round the values: 0.035, 0.070, 0.105, $\cdots$, 0.175. The combined distribution of $M_{ov}/\alpha M_T$ in a number of experiments is said to support this claim.

If this clustering effects exists, it should appear in the distribution of values of M_{ov}/M_T for a sufficiently large number of strictly comparable cells. There should be no need to assume that there is a different value of α for each group of cells, if all cells are similar and are similarly treated. This should remove an unsatifactory feature of the work of MacRobbie[1]. If a single value of influx time is used, it will not be necessary to assume (as MacRobbie[1] seems to have done) that values of M_{ov}/M_T are independent of time at short times. The evidence on this point is far from satisfactory (MacRobbie[2], Fig. 4).

In the present paper we report the results of two large experiments on *Chara corallina*, in each of which the distribution of M_{ov}/M_T was studied. The method of treating the data used by MacRobbie is also critically examined.

MATERIALS AND METHODS

Chara corallina, collected from Ross, Tasmania, is held in culture at Flinders University, South Australia. It is grown in mud and nutrient solution, and frequently replanted. The same species is similarly grown at the University of Sydney, N.S.W., but the starting material was collected at Gledswood, near Sydney. (The Tasmanian material was originally named *C. australis* var. *australis*, the N.S.W. material *C. australis* var. *nobilis*.)

Internodal cells, 3.5–8.0 cm in length, were harvested, prepared as usual for flux experiments and stored some days in Flinders pond water[7]. The cells were pretreated in light in the following solution (Nuffield artificial pond water) for 2 h prior to influx measurement: 2.0 mM *N*-2-hydroxyethylpiperazine *N'*-2'-ethane sulphonic acid[8] (Calbiochem) (HEPES) and 0.7 mM NaOH, 1.0 mM NaCl, 0.1 mM KCl, 0.05 mM $CaSO_4$ (pH 7.0).

The influx solution was Nuffield artificial pond water labelled with $^{36}Cl^-$ (Radiochemical Centre, Amersham) at a high specific activity (7 mC/g Cl^-). This helped to reduce the counting errors for small samples.

After an influx period in Nuffield artificial pond water each cell was rinsed for 15 sec in Flinders pond water and blotted. When drying had reduced its turgor, its end was cut off, and a small drop of vacuolar sap ran out or was gently expressed. A measured volume of this sap, up to half the cell volume, was counted separately from the remainder of the cell. Cell walls and nodal cells were discarded from the remainder before counting. The sum of both counts gave the total radioactivity taken up, while the proportion in the sap could be found from the dimensions of the cell and the assumption that the vacuole occupied 94% of its volume. From these results the calculations gave M_T, M_{ov}, and M_{ov}/M_T.

Radioactivity was measured with a Nuclear–Chicago automatic gas-flow geiger counter. Most samples were counted to 3000 counts, but some to as small a total as 500, of which one-quarter was background. A quantitative estimate of the overall error is given below. No results were discarded due to low counting rate, (*cf.* MacRobbie[1]) since this is not legitimate when the distribution of values is the aim of the experiment.

RESULTS

Expt. A was performed on 99 similar Chara cells from Flinders cultures, during the course of one day. The influx period was 10 min, and the cells were in light at 24°. The scatter diagram obtained by plotting M_{ov} against M_T for these cells is shown in Fig. 1. When the ratio M_{ov}/M_T is plotted against M_T, the scatter diagram of Fig. 2 results. The distribution of these values of M_{ov}/M_T is shown in Fig. 3a. This distribution is clearly not gaussian, and appears bimodal, but there is no appearance of "quantization", *i.e.* of peaks occurring regularly through the distribution.

There is some correlation between M_{ov}/M_T and M_T, as indicated by the results of Table I: the mean influx of cells from the lowest range of values of M_{ov}/M_T is half that of cells from the highest range. No correlation was observed between M_T

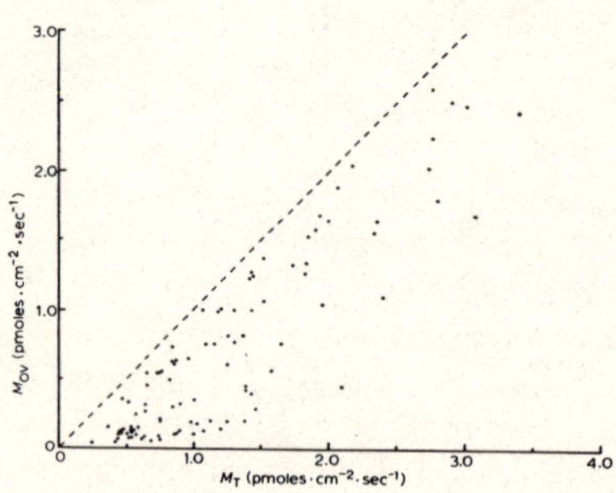

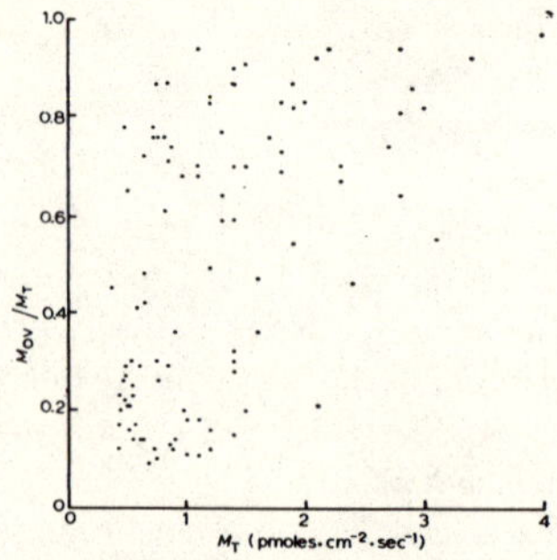

Fig. 1. The influx of Cl^- to the vacuole, M_{ov}, plotted against the influx to the whole cell, M_T, in *Chara corallina*, Expt. A. The influx period was 10 min. The dotted line shows the theoretical upper bound for the values, as the influx to the vacuole cannot exceed that to the whole cell.

Fig. 2. M_{ov}/M_T plotted against M_T, for *Chara corallina*, Expt. A.

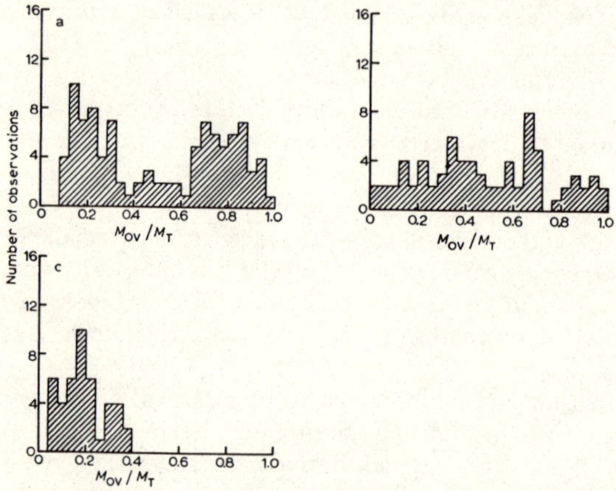

Fig. 3. Histogram showing the distribution of values of M_{ov}/M_T in (a) *Chara corallina*, Expt. A; (*b*) *Chara corallina*, Expt. B; and (c) *Nitella translucens*, Expt. NA3 of MacRobbie[1]. The interval in each histogram is 0.04.

and cell length, and nothing is known that would allow even a rough prediction of M_T or M_{ov} from the size or appearance of a cell.

A similar experiment (Expt. B) done on 74 cells of Chara from N.S.W. by W. J. CRAM, A. W. D. LARKUM AND M. G. PITMAN (personal communication) gave the distribution of M_{ov}/M_T shown in Fig. 3b. The influx period was 6 min. Again there is no sign of regularly spaced peaks, and again a wide scatter of individual values of M_{ov}/M_T.

TABLE I

CORRELATION BETWEEN M_{ov}/M_T AND M_T

Selected range of M_{ov}/M_T	0–0.35	0.35–0.70	0.7–1.0
Number of cells in range	40	22	36
Mean value of M_T	0.82	1.40	1.67
Standard error of mean	0.06	0.16	0.15

DISCUSSION

(a) *The results from Chara corallina*

When a large batch of Chara cells is used in a single experiment with a short influx period, the resulting values of M_{ov}/M_T have a continuous, but not a gaussian, distribution. There is no sign of a discontinuous distribution centred on values in integral ratios, such as would be predicted by the "quantization" of MACROBBIE[1]. Both M_{ov} and M_{ov}/M_T appear to be correlated with M_T at the uptake time of 10 min. MACROBBIE[2] has observed, at short influx times, a correlation of M_{ov} with M_T, and at long times (*e.g.* 60 min) a correlation of M_{ov}/M_T with M_T. However, each of these correlations was observed in the means of batches of 10 cells, differently treated, and not in the scatter of the similarly treated cells of one batch. Thus there is no clear relationship between her observed correlations and ours.

It is to be noted that since M_{ov} and M_T are both positive and since M_{ov} must be less than M_T, a positive correlation of the two quantities must exist *a priori*.

Our actual results are not very dissimilar from those of MACROBBIE[1,2] though our values of M_{ov}/M_T are higher than are found in Nitella. The upper peak in the distribution (Fig. 3a) does not appear in the distribution for her Nitella experiment NA3 (Fig. 3c).

There is no sign of any quantization of values of M_{ov}/M_T in our results.

(b) *The results from Nitella translucens*

The original data of E. A. C. MACROBBIE[1] (personal communication) yield a rather featureless distribution of values of M_{ov}/M_T when they are combined directly without manipulation. Similarly, when the individual experiments are kept separate (Fig. 4), no clustering around integral ratios is apparent. We must therefore examine her manipulations of her data, which lead (i) to correlation coefficients with values near 1.0, and (ii) to histograms which she claimed[1] put the results "in a form in which the quantization is immediately obvious". The following account is based on MACROBBIE[1] and E. A. C. MACROBBIE (personal communication).

For each experiment on 8–42 cells, she ranked the values of M_{ov}/M_T, and chose groups of values, separated by gaps which she says are obvious. Listed below are the values of M_{ov}/M_T in her expts. T and NA3 (Table I, ref. 2). Commas mark the gaps chosen by MacRobbie[1] as separating distinct groups of data values:

Expt. T: 0.16 0.17 0.21 0.21 0.22, 0.32 0.33 0.33 0.33 0.35 0.39 0.45 0.47 0.49, 0.65 0.67 0.67 0.69 0.72

Expt. NA3: 0.046 0.047 0.062 0.066 0.070 0.077, 0.097 0.098 0.107 0.108 0.121 0.124 0.126 0.128 0.141, 0.15 0.16 0:17 0.18 0.18 0.18 0.19 0.19 0.19 0.20 0.20 0.21, 0.22 0.22 0.23 0.24 0.25, 0.28 0.28 0.31 0.32, 0.34 0.34 0.34 0.36 0.38 0.40.

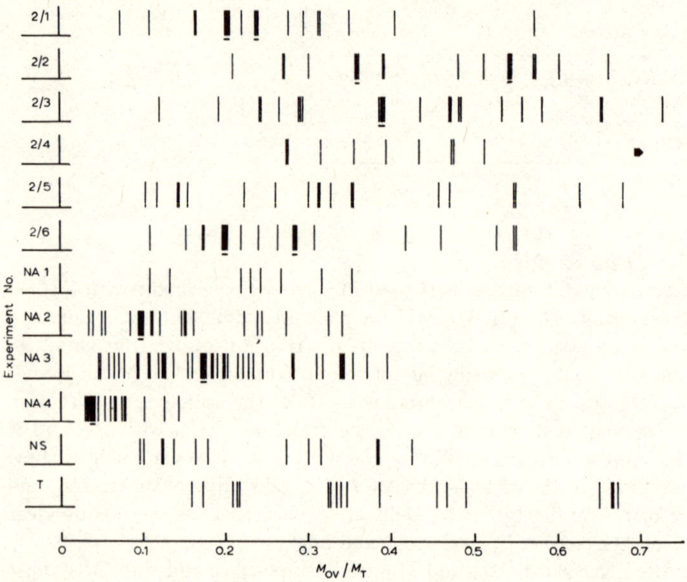

Fig. 4. A distribution diagram of values of M_{ov}/M_T in the experiments of MacRobbie[1]. Each vertical line represents a single value of M_{ov}/M_T. The horizontal bar under some groups of points indicates that all the points in the group have the same value, the mean of the group. The arrow shown in the tabulation for Expt. 2/4 indicates that the following points are off the diagram: 0.909, 0.909, 0.909, 1.07, 1.11.

The mean of the lowest group provided a first estimate of the value of α in the equation $M_{ov} = n\alpha M_T$. All values of M_{ov}/M_T in the list were then divided by this value of α to yield experimental values of $M_{ov}/\alpha M_T$ ($=n'$) and which should approximate to n. This step appears to be necessary because in the Nitella data (Fig. 4) there is often a group of rather low values which is easy to "choose", but no clear grouping among the higher values. The attempt to choose groups among the high values appears to rely heavily on the idea that values of n' must cluster around the integers, *i.e.* in practice the group $n = 3$ in the data consists of those values of n' lying between 2.5 and 3.49. When the allocation of values to groups was complete, a second approximation to α, written $\bar{\alpha}$, was calculated (E. A. C. MacRobbie, personal communication) from the equation

$$\bar{\alpha} = \frac{\Sigma(M_{ov}/M_T)}{\Sigma n x_n}$$

where x_n is the number of values in the nth group. When $\bar{\alpha}$ was obtained, n' was recalculated, and cells were reallocated if necessary to different groups if their value of n' crossed a boundary at 1.5, 2.5 *etc.*, (ref. 1, p. 338). Further iterative steps are possible.

At this point MacRobbie[1] used two procedures: (i) a correlation coefficient was calculated for the linear regression of M_{ov}/n on M_T, and (ii) a histogram was compiled showing the distribution of $M_{ov}/\alpha M_T$ in five or six experiments (about 100 cells in all).

This method of data treatment is open to many objections.

The selection of groups in the data is subjective. Thus Expt. T, listed above, was divided into three groups and $\bar{\alpha}$ was calculated as 0.224 (ref. 1). We have divided it into four groups with a value 0.17 for $\bar{\alpha}$ (Fig. 5). The four groups cluster round the integers 1, 2, 3, 4 just as the three cluster round 1, 2, 3 of MacRobbie[1]. No test known to us will determine which grouping is the "true" one, and we conclude that both are without objective meaning. Data such as that of Expt. NA3 (above) shows no evident grouping, and it may be divided into groups in a variety of ways, each way resulting in a different value of $\bar{\alpha}$. Further, one may impose the same subjective grouping on any list of scattered results: the values of M_T in Expt. T may be grouped with a value of $\alpha = 0.214$ and values of n up to 5.

Since the first steps in the data reduction are subjective rather than objective

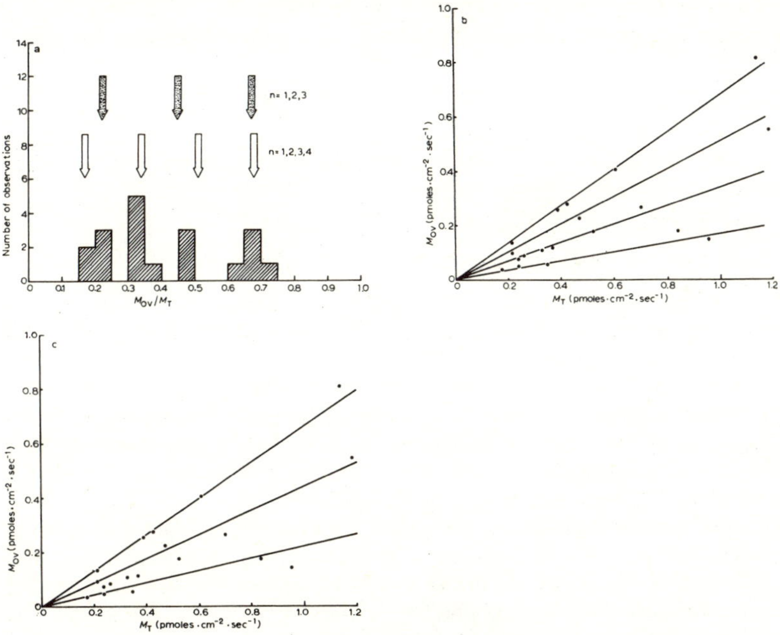

Fig. 5. Aspects of Expt. T with Tolypella[1] showing (a) a histogram of the distribution of values of M_{ov}/M_T with dark arrows indicating the "quantization" of the data of MacRobbie[1], and light arrows, our "quantization". (b) M_{ov} plotted against M_T, with four lines following from our "quantization", drawn through the points. (c) M_{ov} plotted against M_T, with three lines as drawn by MacRobbie[1].

the next steps must be suspect. The calculated regression coefficients for M_{ov}/n on M_T can have no meaning (despite their high values) since the selection of groups based on the slope M_{ov}/M_T automatically produces a good correlation of M_{ov} with M_T within each group. Since the groups are chosen around the integers it follows that a combined regression of M_{ov}/n on M_T will produce high values of r. To check this, normally distributed random values of dummy data of M_{ov} and M_T were produced (IBM Scientific Subroutine package: GAUSS subroutine) and used to calculate values of M_{ov}/M_T. They were grouped as described above, with the following results: G1. Values of M_{ov}/M_T: 0.23 0.26 0.27, 0.39 0.42 0.44 0.54 0.56 0.58 0.62 0.62 0.66, 0.74 0.75 0.80 0.83, 1.00 1.05. $\bar{\alpha} = 0.265$; $r = 0.921$; slope of regression line, 0.250; intercept, 0.011. G2: 0.09 0.24 0.24, 0.29 0.30 0.39 0.40 0.40, 0.48 0.51 0.53 0.61, 0.65 0.69 0.73 0.73, 0.89 0.90, 1.06 $\bar{\alpha} = 0.177$; $r = 0.937$; slope 0.189; (intercept, −0.011. These regression coefficients are as high as some obtained by MacRobbie[1] for her data.

The histograms seem to be open to fewer objections than the regression coefficients, since the values of each experiment are divided only by a single constant $\bar{\alpha}$, before being combined. Since the value of $\bar{\alpha}$ is not uniquely determinable in any one experiment, the shape of the histogram is obviously not unique either, and no conclusions can be drawn from it.

The gap in the histograms at about $n = 1.4$ seems to correspond to a common tendency in the Nitella data for a group of low values to be separate from the higher values. This may be a significant feature of the results, but this would need further tests; it might be a product of experimental technique or represent some real group of cells.

(*d*) *Experimental errors*

An estimate of random experimental errors in such experiments as these can be made. The fluxes M_{ov} and M_T are given by the following expressions, omitting constants:

$$M_{ov} \propto Y_{ss} \cdot V_o \cdot c_o \cdot t^{-1} \cdot Y_o^{-1} \cdot v_{ss}^{-1}$$

$$M_T \propto Y_T \cdot V_o \cdot c_o \cdot t^{-1} \cdot l^{-1} \cdot d^{-1} \cdot Y_o^{-1}$$

where Y_{ss}, Y_T and V_0 are the radioactivities of the sap sample, the whole cell, and the aliquot of solution, respectively, V_{ss} is the volume of the sap sample, V_o is the volume of the aliquot of solution, c_o is the concentration of Cl^- in the solution, t is the influx period, and l, d are the length and diameter of the cell.

All these measurements are subject to errors, the most serious being counting errors in samples of low activity. Counting errors are stated by MacRobbie[1] to have been usually 6–10%, but up to 20%. In the present work the errors in counting the sap samples were less than 7%, and in the other samples 3%.

The total estimated errors are ±8% in M_{ov} and ± 7% in M_T in these experiments, and ± 10% and ± 8% in the experiments of E. A. C. MacRobbie (personal communication). However, to her errors should be added a further amount, not easily estimated, since she has used uptake times varying from 5 to 15 min. The fraction in the vacuole may well vary by some 15% in this time (Fig. 4, ref. 2).

(*e*) *Gaps in data*

The grouping of data of MacRobbie[1] seems to rely on the existence of gaps in the distribution of values in an experiment. No objective criterion exists for the size of a credible gap in experimental data. However, if the expected experimental error is $\pm$ 10% we can scarcely believe in gaps whose size is less than 20% of their median value. On this basis, if we examine her choice of gaps in her data, we find that of 35 gaps chosen to separate data groups, about 19 are greater than or equal to 20% of their median value, and 15 are less. Of gaps in her data not chosen to separate data groups, 9 are greater than or equal to 20% of their median value. This illustrates the subjective nature of the choice of group boundaries, and suggests again that it was biased towards the production of groups round the integers.

CONCLUSIONS

"Quantization" is hardly a feature of the data which leaps to the eye of the observer. It must be considered as an hypothesis subject to experimental test.

The grounds on which hypotheses for test are normally set up appear to be two: (i) that, in accordance with Occam, they are the simplest unfalsified hypotheses available, or (ii) that they follow from some reasonable argument on *a priori* grounds. Such an hypothesis may win some acceptance when tests which might well have falsified it can be shown to have failed to do so.

Looking at the quantization hypothesis in these terms one concludes, first, that it is far from being the simplest available, second, that no argument known to us suggests it.

There seems no reason to accept its existence as a phenomenon or its usefulness as an hypothesis. If its usefulness as an hypothesis were accepted, it would appear that its testing would require much more evidence than is available to date, or more sophisticated tests.

It does seem clear that there may be unexpected features in the distribution of values of M_{ov}/M_T. No further conclusion seems justified at the present.

ACKNOWLEDGEMENTS

This work was supported by grants from the Australian Research Grants Committee and from the Nuffield Organization (Australia) for which we are grateful.

We thank Dr. E. A. C. MacRobbie for discussing her results with us, for providing us with a copy of her paper before publication, and for permission to quote her experimental data. This does not imply her agreement with our point of view.

REFERENCES

1 E. A. C. MacRobbie, *J. Exptl. Botany*, 21 (1970) 335.
2 E. A. C. MacRobbie, *J. Exptl. Botany*, 20 (1969) 236.
3 E. A. C. MacRobbie and J. Dainty, *J. Gen. Physiol.*, 42 (1958) 335.
4 J. M. Diamond and A. K. Solomon, *J. Gen. Physiol.*, 42 (1959) 1105.
5 A. W. D. Larkum, *Nature*, 218 (1968) 447.
6 A. W. D. Larkum and M. Pring, in preparation.
7 G. P. Findlay, A. B. Hope, M. G. Pitman, F. A. Smith and N. A. Walker, *Biochim. Biophys. Acta*, 183 (1969) 565.
8 N. E. Good, G. D. Winget, W. Winter, T. N. Connolly, S. Izawa and R. M. M. Singh, *Biochemistry*, 5 (1966) 467.

CHARACTERIZATION OF A FLAVODOXIN FROM THE GREEN ALGA CHLORELLA

Walter G. Zumft[+] and Hart Spiller

SUMMARY

A new flavoprotein has been isolated and purified to homogeneity from the green alga *Chlorella fusca*. It has characteristics comparable to those of phytoflavin from the blue-green alga *Anacystis nidulans* or flavodoxin from various bacteria. Chlorella flavodoxin can replace spinach and *Chlorella* ferredoxin in photosynthetic NADP reduction of broken chloroplasts. Its molecular weight is 22 000. One mole of FMN is present per mole of protein. The extinction coefficient at 464 nm is 10 000 M^{-1} cm^{-1}. Reduction of the flavoprotein proceeds via a blue flavosemiquinone radical. The amino acid composition reveals a close relationship to bacterial flavodoxins.

Low molecular, low potential flavoproteins as a new class of electron carriers are capable of replacing ferredoxin in its various reactions (1). Smillie was the first to isolate from the blue-green alga *Anacystis nidulans* a flavoprotein, named phytoflavin that acts instead of ferredoxin in the light-dependent reduction of NADP by chloroplasts (2). Later on closely related flavoproteins, flavodoxins, were obtained from various bacteria (1). Recently Vetter and Knappe reported the isolation of a ferredoxin-like substance as well as a flavodoxin from the bacterium *Escherichia coli* K 12 (3).

The properties and functional role of phytoflavin from *Anacystis* in a series of photosynthetic reactions have been studied by Bothe (4) and Bothe et al. (5). Since flavodoxins have been found only in procaryotic organisms they seemed to be a unique feature of bacteria and blue-green algae. We wish to report on the detection of a flavodoxin[+] from the eucaryotic green alga *Chlorella*. Similar investigations with *Ankistrodesmus braunii* under virtually the same conditions yielded another flavoprotein mediating NADP reduction of broken chloroplasts. Thus an ubiquituous distribution of this electron transfer protein seems to be probable. Some of its properties have been investigated and are compared with those of bacterial flavodoxins, ferredoxin, and phytoflavin from other sources.

MATERIAL AND METHODS

Cells of *Chlorella fusca* Shihira et Krauss, strain 211-15 (=pyrenoidosa), from the culture collection at Göttingen, were grown as described elsewhere (6). Instead of 8 mg $FeSO_4.2H_2O$/l only 0.03 mg were used to produce iron-deficient conditions. This concentration still allowed a good growth of the organism giving at the same time high yields of the electron carrier. After 12 days the cells were broken and the resulting homogenate was freed from cell particles by high-speed centrifugation. The purification procedure of flavodoxin resembled that of ferredoxins and consisted of chromatography on DEAE-cellulose (DE-52, Whatman), elimination of nucleic acids by precipitation with a neutral 2 % protamine sulfate solution, a second passage through DEAE-cellulose and gelfiltration in Sephadex G-75. This process gave a homogeneous protein as shown by analytical disc electrophoresis in a 7.5% gel at pH 6.0. The specific activity, measured as micromoles NADP reduced/min per mg protein rose from 0.015 in the centrifugation supernatant to 2.83 after the last purification step, resulting in an overall purification of approx. 200-fold. The activity of flavodoxin was assayed like ferredoxin (7). Broken chloroplasts (8) and ferredoxin (9) were prepared from green-house spinach. *Chlorella* ferredoxin was obtained by a method of Zumft (10). Gel filtration experiments were performed with a Sephadex G-75 column (1.5 x 80 cm) according to Andrews (11).

Polyacrylamide electrophoresis in the presence of dodecylsulfate using the normal amount of cross linker followed the method

[+]Following the suggestions of Bothe, Hemmerich and Sund (5) one of the names should be given up. Despite Smillie's nomenclature we prefer the more relevant name 'flavodoxin'.

of Weber and Osborn (12). FMN was identified by paper chromatography (13); extinction coefficients were determined by dissociation of FMN from the protein by trichloracetic acid (14). Spectral measurements were done in a Zeiss PMQ II or a Beckman DB-GT spectrophotometer.

RESULTS AND DISCUSSION

Flavodoxins are distinguished from other flavoproteins by their low redox potential and low molecular weight of approx. 15 000. They are synthesized well only under iron deficiency, contain FMN and can replace ferredoxin in photosynthetic NADP reduction of broken chloroplasts (1). Therefore this assay system was used to identify flavodoxin in our preparations, also obtained from low-iron cultures. Fig. 1 shows that flavodoxin from _Chlorella_ in equimolar amounts is 60% as active as spinach or _Chlorella_ ferredoxin. The replacement of ferredoxin by flavodoxin in vitro is amply matched by the occurence of both in vivo. Our results show that in cells (1 kg wet weight) from a complete culture medium (6) flavodoxin occurs beside ferredoxin in a molar ratio of 1:10. On the other hand under

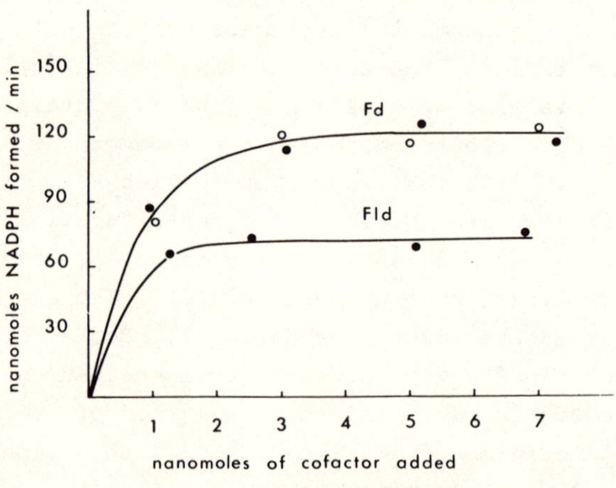

Fig. 1 Flavodoxin (Fld)- or ferredoxin (Fd)-catalyzed NADP reduction by spinach chloroplasts. A 1 cm cuvette contained in a volume of 2 ml 0.6 micromoles NADP, 100 micromoles Tris (pH 7.5), 20 micromoles $MgCl_2$, 0.1 mg chlorophyll and cofactors as indicated. Reference cuvette without cofactor; light intensity 30 000 lux. Upper curve ferredoxin from spinach or _Chlorella_, lower curve flavodoxin from _Chlorella_.

severely iron-deficient conditions no or only trace amounts of ferredoxin were detected, whereas appreciable quantities of cytochromes and the iron protein nitrite reductase (10,15) still were found. This points to a certain order of preference according to vital importance of iron proteins.

Fig. 2a shows the absorption spectrum of oxidized flavoprotein with absorption maxima at 275, 379 and 464 nm, respectively. Small shoulders were found at 292 and 490 nm. Characteristic absorption ratios are as follows, $A_{275\ nm}/A_{379\ nm} = 6.04$ and $A_{275\ nm}/A_{464\ nm} = 5.46$. The millimolar extinction coefficient at 464 nm is 10 ± 0.2. The spectrum is typical for protein-bound flavin with a remarkable shift of the main flavin peak over approx. 20 nm. *Anacystis* phytoflavin exhibits the same absorption maxima (2,4) whereas the red absorption band of bacterial flavodoxins has a small blue shift (13,14). Under nitrogen atmosphere in the presence of 50 mM EDTA and Na-acetate (pH 5.6), and on irradiation with a tungsten actinic light source flavodoxin is slowly but reversibly reduced to the grey-blue semiquinone form (Fig. 2b). On raising the pH to 8 and further addition of formamidine sulfinic acid in 200-fold excess one can gradually observe the decrease of the semiquinone absorption and the appearance of the fully reduced form (Fig. 2c). Semiquinone maxima

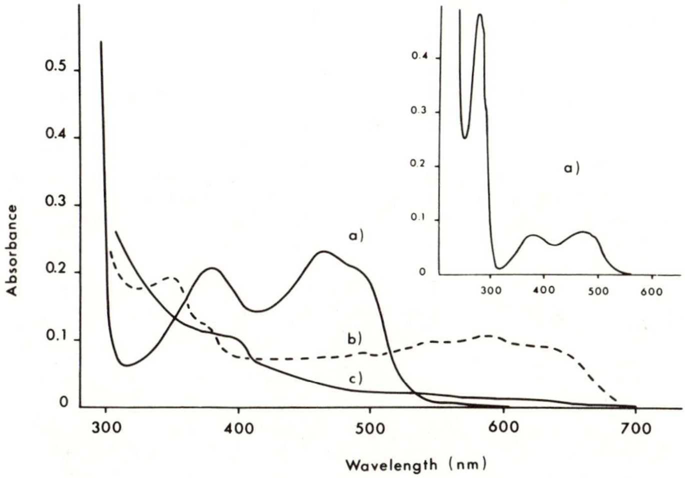

Fig. 2 Absorption spectrum of flavodoxin from *Chlorella*. Curve a) oxidized form, b) semiquinone form, c) reduced form; for details see the text.

are at 350 and 588 nm. Isosbestic points of the couple oxidized flavodoxin/semiquinone are at 359 and 517 nm. Isosbestic points of the couple fully-reduced flavodoxin/semiquinone are at 329 and 385 nm. Furthermore the fully reduced form shows a maximum at 395 nm. Readmittance of oxygen reverses the reduction. These observations are similar to those made by Massey and Palmer (16) and Mayhew and Massey with bacterial flavodoxin (14).

By using the electrophoretic technique in polyacrylamide with trypsin inhibitor (soybean), myoglobin, hemoglobin, ribonuclease and chymotrypsin we determined a molecular weight of 22 000 (see Fig.3). Besides, we found it rather surprising to get with this method even in most thoroughly purified preparations another band of just half the molecular weight. Experiments are in progress to reveal if this is due to a splitting of the molecule under those strongly dissociating conditions or an unusual behavior of Chlorella flavodoxin during the dodecylsulfate electrophoresis.

By gel filtration studies in Sephadex G-75 and using myoglobin, bovine serum albumin, chymotroypsinogen A and cytochrome c as standards we estimated an average molecular weight of 21 000. Amino acid composition indicated a molecular weight of

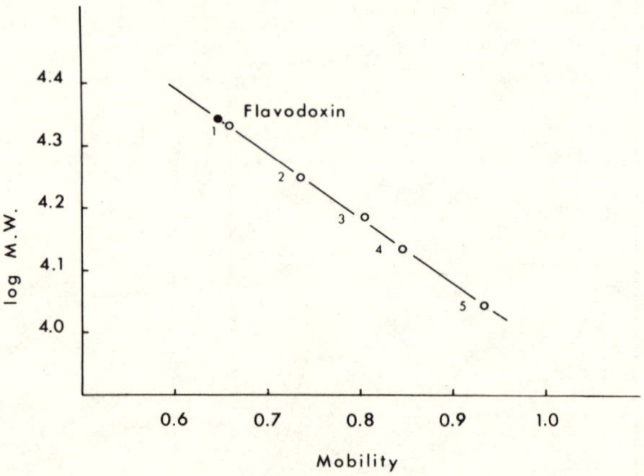

Fig. 3 Molecular weight determination of flavodoxin by dodecylsulfate electrophoresis. In each case 20 micrograms of protein were applied to the gel tube. Protein standards (1) trypsin inhibitor (soybean), (2) myoglobin, (3) hemoglobin, (4) ribonuclease and (5) chymotrypsin (11 000 chain plotted) were obtained from Serva (Heidelberg).

21 700. From these data we conclude a final molecular weight of 22 000 daltons.

The identification of the prosthetic group by paper chromatography disclosed FMN and not FAD as part of the protein. This is in agreement with findings on *Anacystis* phytoflavin (2) and bacterial flavodoxins (1). The quantitative estimation revealed a content of 1.1 µg FMN per 63 µg of protein. This corresponds to 1 FMN per 27 000 g of protein and suggests the binding of one FMN molecule per protein molecule.

The amino acid composition is shown in table I. According to the higher molecular weight in comparison with bacterial flavodoxins the molecule consists of 206 amino acids with a prevalence of certain amino acids such as aspartic and glutamic acid glycine and alanine. As regards comparable amino acids there is a good resemblance with those of bacterial flavodoxins.

In conclusion, the *Chlorella* flavoprotein seems to be a new member in the class of flavodoxins by its absorption spectrum, formation of the blue radical upon reduction, similarity of the

TABLE I

AMINO ACID COMPOSITION OF FLAVODOXINS

	Chlorella fusca[a)]	*Clostridium pasteurianum* (13)
	(residues per molecule)	
Lys	10	10
His	4	0
Arg	4	2
Asp	27	18
Thr	12	4
Ser	12	14
Glu	24	19
Pro	5	4
Gly	25	17
Ala	24	14
Half-cystine	5	1
Val	13	15
Met	1	4
Ile	8	5
Leu	18	12
Tyr	6	2
Phe	4	3
Trp[b)]	4	4
Total	206	148

[a)]calculated values from 20 and 70 h hydrolysis,
[b)]determined by hydrolysis with 2% thioglycolic acid (17).

amino acid composition and identity of the prosthetic group. It is only different in its slightly higher molecular weight. Further physico-chemical properties of the flavodoxin from Chlorella and some aspects of its physiological role in nitrate assimilation will be presented in a subsequent paper.

ACKNOWLEDGEMENTS

We wish to thank Prof. E. Kessler for his support during this work. Mrs. Manuela Chaves joined parts of this research. For reliable assistance we feel greately indebted. We thank also Dr. L. Gürtler for the amino acid analysis. Mr. H. Wurzbacher from the botanic garden kindly supplied us with spinach. We acknowledge the help of the Deutsche Forschungsgemeinschaft and the Stiftung Volkswagenwerk.

REFERENCES

(1) Neims, A.H., L. Hellermann, Ann.Rev.Biochem., 39, 867 (1970)

(2) Smillie, R.M., Plant Physiol., 40, 1124 (1965)

(3) Vetter, H.Jr., J. Knappe, Hoppe-Seyler's Z.Physiol. Chem. 352, 433 (1971)

(4) Bothe, H., in: H. Metzner, ed., Progress in Photosynthesis Research, vol. 3, p. 1483; H. Laupp, Jr., Tübingen (1969)

(5) Bothe, H., P. Hemmerich, H. Sund, in: Flavins and Flavoproteins, Proc.3rd Conf. Flavins and Flavoproteins, Durham, 1969 (in the press)

(6) Kessler, E., F.-C. Czygan, Arch.Mikrobiol., 70, 211 (1970)

(7) San Pietro, A., in: S.P. Colowick and N.O. Kaplan, eds., Methods in Enzymology, vol. 6, p. 499, Academic Press, New York (1963)

(8) Whatley, F.R., D.I. Arnon, in: S.P. Colowick and N.O. Kaplan, eds., Methods in Enzymology, vol. 6, p. 308, Academic Press, New York (1963)

(9) Tagawa, K., D.I. Arnon, Nature (Lond.) 195, 537 (1962)

(10) Zumft, W.G. unpublished, manuscript in preparation

(11) Andrews, P., Biochem., J., 91, 222 (1964)

(12) Weber, K., M. Osborn, J.Biol.Chem. 244, 4406 (1969)

(13) Knight, E. Jr., R.W.F. Hardy, J.Biol.Chem. 241, 2752 (1966)

(14) Mayhew, S.G., V. Massey, J.Biol.Chem., 244, 794 (1969)

(15) Aparicio, P.J., J. Cárdenas, W.G. Zumft, J.Mª. Vega, J. Herrera, A. Paneque, M. Losada, Phytochemistry (in the press)

(16) Massey, V., G. Palmer, Biochemistry, 5, 3181 (1966)

(17) Matsubara, H., R.H. Sasaki, Biochem.Biophys.Res.Commun., 35, 175 (1969).

OBSERVATIONS ON PHOTOSYSTEM II MUTANTS OF SCENEDESMUS: PIGMENTS AND PROTEINACEOUS COMPONENTS OF THE CHLOROPLASTS*

NORMAN I. BISHOP AND JAMES WONG

SUMMARY

Nine mutants of the green alga, *Scenedesmus obliquus*, which are blocked in the Photosystem II portion of photosynthesis were analyzed for possible deletion or alteration of (1) various components of the photosynthetic electron transport system, (2) of chloroplast lipids, (3) of total chlorophyll or of the chlorophyll *a*/chlorophyll *b* ratio, and (4) of their content of carotenes and carotenoids. No changes in content or activity of ferredoxin, ferredoxin–NADP$^+$ reductase, plastocyanin, cytochrome *c*-552, and the membrane-bound *b*-type or *c*-type cytochromes were observed. The most consistent differences noted between the mutant strains and the wild-type strain were in the molar ratio of chlorophyll/plastoquinone A, the total chlorophyll content, and a decreased content of α- and β-carotene with a concomitant increase of carotenoids. The loss of Photosystem II activity in these mutant strains, as observed either with whole cells or with isolated chloroplast fragments, may be accounted for by their decreased content of plastoquinone A. Their decreased chlorophyll content and altered carotene/xanthophyll ratio also suggests possible alteration of chloroplast membranes resulting in increased internal oxidation of the photosynthetic pigments.

INTRODUCTION

Within recent years, a number of mutations of a variety of photosynthetic organisms have been employed to study specific aspects of photosynthesis such as the mechanism of the formation of chloroplast pigments[1–3], the details of chloroplast development[4–6], and the role of certain pigments in different partial reactions of photosynthesis[7–10]. Specific studies on the mechanism of function of Photosystem I and Photosystem II with mutations have been limited largely to the green algae, *Chlamydomonas reinhardi*[11–13] and *Scenedesmus obliquus*[14–16]. In these algae photosynthetic mutants have been isolated wherein the activity of Photosystem II has been specifically suppressed. In more detailed studies in this laboratory, and in cooperation with other research groups, it has been shown that although a normal

Abbreviations: DCMU, 3(3,4-dichlorophenyl)-1,1-dimethylurea; PQ-A, plastoquinone A; DCIP, 2,6-dichlorophenolindophenol.

* A portion of the results presented here were reported at the Biophysics Meeting, Los Angeles, Calif., 1969.

Photosystem I reaction is extant in these Scenedesmus mutants, no evidence for Photosystem II activity has been observed. These observations include the following: the light-initiated EPR signal has only the characteristics of the fast signal[17]; isolated chloroplasts only show reactions typical of Photosystem I[16]; certain light-induced absorbance changes show principally the function of the long-wavelength photoreactions[16–18]; and no characteristic delayed light emission typical of the normal phenotype has been detected[19]. Also, derivative absorption spectroscopy (at room and liquid nitrogen temperatures) revealed that no major alteration has occurred in the *in vivo* forms of chlorophyll[21] and more recent studies with the electron microscope have revealed no conspicuous alteration of chloroplast structure (N. I. BISHOP, unpublished data).

In attempts to explore further the nature of the mutation which results in the loss of Photosystem II activity, we have examined more closely the various chloroplast pigments and a number of the proteinaceous components of the chloroplasts. It will be reported that in nine individual Photosystem II mutants, the most consistent differences noted between them and the normal phenotype of Scenedesmus are a decrease in the total amounts of chlorophylls and of plastoquinone A synthesized.

METHODS AND MATERIALS

Culture and maintenance of algal strains

Normal wild-type *Scenedesmus obliquus* (strain D_3) and various photosynthetic mutants were grown heterotrophically on a basal nitrate medium[16] which was supplemented with glucose (0.5 %) and yeast extract (0.25 %). Screw-topped 500-ml erlenmeyer flasks, each containing 250 ml of culture medium, were inoculated with the appropriate algal strain, placed on a thermostated rotary shaker[28], and grown for a maximum of 3 days. Myxotrophic cultures were grown on the same medium, but in culture tubes in an illuminated growth chamber as previously described[21]. The packed cell volume of each culture was determined by centrifugation in "protein constable tubes" (Cenco Scientific). From such determinations it was possible to obtain equal cell volumes of each of the mutants and of the wild type for comparison studies.

Isolation and purification of algal plastoquinone

Approx. 10–15 g wet weight of washed cells were extracted 4 times with 30-ml portions of hot methanol. The extracts were combined in a 250-ml volumetric flask and brought to volume. An aliquot of this extract was used to determine the total chlorophyll according to the method outlined by HOLDEN[22]. Chlorophyll *a* and *b* amounts were determined from the appropriate extinction coefficients obtained by MACKINNEY[23].

The methanol extract was evaporated to dryness and washed onto a 2.5 cm × 15 cm column of 100 mesh silicic acid with two 50-ml washes of isooctane (2,2,5-trimethylpentane). Upon adsorption of the extract, the β-carotene was eluted from the column with 125 ml of $CHCl_3$–isooctane (15:85, v/v). Subsequent elution with 150 ml $CHCl_3$–isooctane (75:25, v/v) removed the plastoquinones. The column chromatography procedure was expedited by using a water-aspirator vacuum and collecting the fractions in 250-ml suction flasks. The plastoquinone fraction was evaporated to dryness for further purification by thin-layer chromatography.

Purification of the plastoquinone was achieved by a 3-step thin-layer chromatography procedure: (1) The extract was taken up in chloroform, streaked onto a 20 cm × 20 cm silica gel plate and developed in a benzene–heptane solvent (60:40, v/v ($R_f = 0.45$). (2) The plastoquinone was scraped off, eluted, streaked on a paraffin treated silica gel plate (immersed in a 5 % solution of liquid paraffin in ligroine, b.p. 66–75°) and developed in 3 % H_2O-acetone solvent ($R_f = 0.38$). The plastoquinone was located on the plate as a purple band when the plate was sprayed with a 0.1 % rhodamine in absolute ethanol solution and viewed under ultraviolet light. (3) The plastoquinone was eluted free of the dye and the silica gel with redistilled chloroform, restreaked onto a 5 cm × 20 cm silica gel plate and developed in redistilled petroleum ether (b.p. 36–43°) until all the paraffin was removed from the origin. The plastoquinone, which did not move from the origin, was removed from the silica gel with chloroform and then evaporated to dryness. The purified sample was taken up in buffered ethanol (0.01 M ammonium acetate, pH 5.0, in absolute ethanol) for spectrophotometric analysis.

The $\Delta E^{1\%}_{1\,\mathrm{cm}}$ of 200 (at 255 nm) for the reduced–oxidized spectrum of plastoquinone was used to calculate the plastoquinone concentration of each of the mutant strains[24]. From this calculation, molar concentrations were estimated and a molar ratio of quinone to chlorophyll determined. For such calculations, an average molecular weight for chlorophyll was calculated based upon the ratio of chlorophyll *a* to chlorophyll *b* in the samples, which was 3.4 ± 0.4 in wild type cells.

The method described for the purification of plastoquinone effectively separates all of the algal quinones, including the various plastoquinones, vitamin K, and tocopherylquinones. Moreover, the plastoquinone is purified from a number of components which interfere with its determination; these components are most likely a variety of fatty acids which also have ultraviolet absorption and show alteration of spectra upon treatment with borohydride. When pure authentic plastoquinone A was run through the complete purification procedure, 75 % of the original amount was recovered.

Preparation and purification of chloroplast proteins

For studies on the electron transport components of the chloroplast, large-scale cultures were grown in 3-l Fernbach flasks; a total of 22 l of culture medium was required to provide about 120 g fresh weight of packed cells. The cells were washed with 0.2 M Tris–HCl (pH 7.5) and treated according to the method employed by Powls *et al.*[25]. Further purification of ferredoxin–NADP$^+$ reductase, ferredoxin, plastocyanin, and cytochrome 552 was secured by chromatography on DE-32, Sephadex G-100, and hydroxylapatite according to the procedures of Gorman and Levine[26,27] and Matsubara[28]. The activity of each of the components, except for the cytochrome *c*-552, was determined with isolated spinach chloroplasts that had been freed from soluble proteins by extraction with 0.8 M Tris (pH 8.0). This treatment was essentially the same as the reported by Yamashita and Butler[29] except that the extraction procedure was continued for 3 h at 4°. Since this procedure effectively removes any Photosystem II activity, either the electron donor system dichlorophenolindophenol–ascorbate or diphenylcarbazide[30], was employed in analysis of NADP$^+$ photoreduction. The activity of reduced cytochrome *c*-552 as an electron donor in a photochemical system was tested by measuring the rate of its photooxidation by a digitonized fraction of spinach chloroplast[31].

Assay of membrane-bound components

The isolation of insoluble chloroplast components and anaylsis for bound cytochrome *c*-552, cytochrome *b*-559 and cytochrome *b*-562 were performed according to the method previously published[25].

Lipid analysis

1 ml of packed cells of each of the phenotypes was extracted with 50 ml of hot methanol, centrifuged at 2000 × *g*, the pellet extracted with 50 ml of methanol–chloroform (1:1, v/v), and centrifuged. The pellet from this step was extracted once more with chloroform, centrifuged and the entire extracts combined and evaporated to dryness.

The various lipids and pigments contained in the extract were separated by either one or two-dimensional thin-layer chromatography on Silica Gel G. For one-dimensional analysis, 20 cm × 20 cm plates were spotted with 0.2 ml of extract and run in a solvent consisting of chloroform–methanol–water (65:25:4, by vol.). For two-dimensional analysis, the first solvent was the same as above and the second solvent consisted of *n*-butanol–acetic acid–water (40:25:5, by vol.). Lipid-like components on the chromatograms were detected by exposing the plates to iodine vapor or by spraying the plates with 0.1 % rhodamine B (in ethanol) and viewing the quenched fluorescence under ultraviolet light.

Further characterization of the lipid components was performed by growing the algae in the presence of either $NaH_2{}^{32}PO_4$ or uniformly ^{14}C-labeled glucose prior to the extraction and chromatography. Labeled substrates were added during the logarithmic growth phase for a period of 4 h and the cells were subsequently harvested and extracted as indicated above. After chromatographic separation of the various components, the thin-layer chromatography plates were exposed to "No-Screen" X-ray film for a period of 7 days for the ^{32}P-labeled extracts and for 14 days for the ^{14}C-labeled extracts. The thin-layer chromatography plates were then sprayed with rhodamine B and compared to the patterns observed on the developed X-ray film.

Carotene and carotenoid analysis

For a quantitative analysis of the total complement of carotenoids in mixotrophic or heterotrophically grown cells of Scenedesmus, it was essential to select cells that were in the logarithmic growth phase, which was obtained by inoculating fresh medium with 50 μl of cells and harvesting after 2 days of growth. 1 ml of packed cells was extracted 3 times with 50 ml portions of hot methanol and the combined extract saponified by the addition of 20 ml of 5 M NaOH. The saponification was accelerated by heating the extract to boiling and then quickly cooling to room temperature.

The saponified extract was diluted to approx. 80 % methanol with water and the carotenoids were partitioned into petroleum ether. The partitioning was expedited by the addition of 10 % NaCl; partitioning with petroleum ether was repeated until all carotenoids had been removed from the lower phase. The petroleum ether fractions were then combined, washed with water, dried with anhydrous Na_2SO_4, evaporated to dryness under vacuum and redissolved in 5 ml of chloroform.

1 ml of the extract was streaked onto 20 cm × 20 cm thin-layer chromatography plates of Silica Gel G and developed in a solvent composed of petroleum ether–

isopropanol–water (100:10:0.5, by vol.). Prior to development, the plates were preincubated with this solvent in order to obtain better resolution of the several carotenoid bands. After development, individual bands were removed from the plates and extracted from the silica gel with acetone. Individual samples were evaporated to dryness and the appropriate solvent added for the spectrophotometric determination of the concentration of each of the components. Although some 10–11 separate compounds or bands can be separated by this procedure, only those values for $\alpha + \beta$-carotene and lutein + zeaxanthin will be reported in this paper.

The procedure outlined above does not separate the isomeric forms of carotene (α- and β-carotene) and carotenoid (lutein and zeaxanthin). Their separation was obtained by rechromatographing the bands containing the isomers on thin-layer chromatography plates on an adsorbent consisting of $CaCO_3$–$Ca(OH)_2$–MgO (29.5:5.0:6.0, by weight). Effective separation of α- and β-carotene was obtained with a solvent consisting of 1 % isopropanol in isooctane; by increasing the isopropanol concentration to 10 %, adequate separation of lutein and zeaxanthin was obtained.

The amounts of carotene and carotenoids contained in each of the samples were determined spectrophotometrically by employing the $E^{1\%}_{1\,cm}$ values for the predominant isomer, *i.e.* β-carotene and lutein, in either petroleum ether or in ethanol respectively[22].

RESULTS AND DISCUSSION

General characteristics of mutant strains employed

For this study, nine of the mutants of Scenedesmus in our collection which have the general characteristics required of a Photosystem II mutant were examined. Three other mutant types were used for comparison against the Photosystem II mutants as well as against the wild-type strain. The additional strains employed were ScD_3-8, ScD_3-26, ScD_3-50. The characteristics of each of these strains has been described in part elsewhere. In particular, ScD_3-8 is deficient in the reaction center chlorophyll of Photosystem I, *i.e.* it has no detectable P-700[17,20]; ScD_3-26 lacks membrane-bound cytochromes[25], has poorly developed chloroplasts, and appears to lack chloroplast ribosomes; ScD_3-50 lacks cytochrome *c*-552[25]. None of these mutants possesses, the *in vivo* capacity for photoreduction in contrast to the Photosystem II mutants but through chloroplast studies and fluorescence analyses it is known that they possess some Photosystem II activity. In Table I, the general *in vivo* reaction characteristics of the mutant forms employed in this study are compared to the wild-type Scenedesmus. In Table II, the ability of chloroplasts isolated from the various strains to carry out the photoreduction of either $NADP^+$, of cytochrome *c*, or of DCIP (with electrons from water or from 2,5-diphenylcarbazide) or of $NADP^+$ with DCIP–ascorbate as the electron donor system are summarized. Consideration of the data in Tables I and II reveals the clear distinction between the Photosystem II mutants and the other types used for comparison. The inability of chloroplasts isolated from the Photosystem II mutants to photoreduce DCIP (with diphenylcarbazide as the electron donor) indicates that all of these mutants lack the reducing end of Photosystem II.

Variations in the capacity for photophosphorylation in some of these mutant strains have been discussed previously[16].

TABLE I

THE PHOTOCHEMICAL ACTIVITY OF WHOLE CELLS OF VARIOUS MUTATIONS OF *Scenedesmus obliquus* (HETEROTROPHIC)

Photosynthesis, the quinone-Hill reaction, and photoreduction were measured manometrically employing the all-glass differential respirometer (Gilson Medical Electronics). Photosynthesis measurements were made in No. 9 carbonate–bicarbonate buffer pre-equilibrated with 1 % CO_2–air. Freshly sublimed *p*-benzoquinone was retained in the sidearm of a reaction flask prior to initiating the Hill reaction which was performed in 96 % N_2–4 % CO_2. Photoreduction was determined in cells that had been adapted for 6 h to an atmosphere of 96 % H_2–4 % CO_2; for the measurement of photoreduction in the wild type DCMU, at a final concentration of $4 \cdot 10^{-6}$ M, was added prior to initiating rate measurements. All manometric experiments were done with freshly harvested cells which were selected from cultures 50–56 h old. The variable yield fluorescence data represent the augmentation of 686 nm fluorescence emission (as excited by 436-nm wavelength light) caused by the addition of a supplemental 650-nm wavelength actinic light. Further considerations on the measurements of fluorescence are included in the following paper.

Algal strain	*Photosynthesis (μl O_2/h per mg chlorophyll)*	*Hill reaction (μl O_2/h per mg chlorophyll)*	*Photoreduction (μl CO_2/h per mg chlorophyll)*	*Variable yield fluorescence (% increase)*
ScD_3 wild-type	880	530	378	20
Photosystem II mutants				
ScD_3-a′	11	15	400	1
ScD_3-4	24	25	387	0
ScD_3-5	20	25	350	1
ScD_3-10	20	17	360	0
ScD_3-11	16	20	411	0
ScD_3-15	62	30	366	0
ScD_3-40	16	23	397	1
ScD_3-42	10	13	430	0
ScD_3-67	54	38	342	3
Photosystem I or electron transport mutants				
ScD_3-8	25	280	13	110
ScD_3-26	0	20	0	57
ScD_3-50	0	23	0	74

Pigment analyses

In earlier publications concerning some of the general features of the Photosystem II mutants used in this study, it was reported that no major change in pigmentation was observed. The more detailed analysis that has been made recently revealed a minor but consistent difference between the Photosystem II mutants and the wild-type strain. The most significant difference noted in the analysis of the chlorophylls was an approx. 25 % decrease in the total chlorophyll in all of the Photosystem II mutants. A similar decrease was not seen in the other mutants studied. Although total chlorophyll is apparently decreased, and most strikingly in mutant 42, no significant difference in the production of chlorophyll *a* over that of chlorophyll *b* was observed, *i.e.* the ratio of these chlorophylls in the mutants remained essentially the same as in the wild type. These data are summarized in Table III along with other information on the carotene content of the mutants.

From the data of this table, it is also apparent that the Photosystem II mutants accumulate a lesser amount of α- + β-carotene and lutein + zeaxanthin than the wild-type. Additional information (to be published elsewhere) will show that the

TABLE II

THE PHOTOCHEMICAL ACTIVITY OF ISOLATED CHLOROPLAST FRAGMENTS OF WILD-TYPE AND VARIOUS MUTATIONS OF *Scenedesmus obliquus* (HETEROTROPHIC)

The methodology employed for the measurement of the algal chloroplast reactions summarized below was essentially the same as described by PRATT AND BISHOP[16] except that younger cells were employed (see METHODS). Measurements of the reduction of DCIP with 2,5-diphenylcarbazide as the hydrogen donor were performed according to the procedure of VERNON AND SHAW[30]. Photosystem II activity of the wild-type chloroplast was inactivated by heating for 10 min at 50° prior to measuring the photoreduction of DCIP with diphenylcarbazide (DCP) as the hydrogen donor.

Chloroplast source	*Reaction mixture*			
	DCIP	*NADP+*	*DCIP–ascorbate–NADP+*	*DCIP–DPC*
ScD_3-wild-type	98	238	141	92
ScD_3-a′	0	5	128	0
ScD_3-4	1	5	130	0
ScD_3-5	1	5	110	2
ScD_3-10	2	5	98	3
ScD_3-11	0	5	146	0
ScD_3-15	1	5	137	2
ScD_3-40	3	5	128	0
ScD_3-42	4	5	110	10
ScD_3-67	6	5	140	4
ScD_3-8	46	5	5	88
ScD_3-26	3	5	10	2
ScD_3-50	1	5	78	2

TABLE III

PIGMENT CONTENT OF WILD-TYPE AND VARIOUS MUTANT FORMS OF *Scenedesmus obliquus* (HETEROTROPHIC)

Algal strain	*Total chlorophyll (mg)*	*Chlorophyll a (mg)*	*Chlorophyll b (mg)*	*Chl a / Chl b*	*α- + β-Carotene (mg)*	*Lutein + zeaxanthin (mg)*
Wild-type	5.90 ± 0.78	4.56 ± 0.54	1.34 ± 0.22	3.40 ± 0.43	0.12 ± 0.02	0.23 ± 0.03
System II mutants						
ScD_3-a′	4.88 ± 0.44	3.80 ± 0.30	1.07 ± 0.27	3.74 ± 0.79	0.07 ± 0.01	0.17 ± 0.02
ScD_3-4	4.54 ± 0.74	3.65 ± 0.58	0.90 ± 0.17	4.11 ± 0.40	0.09 ± 0.02	0.18 ± 0.02
ScD_3-5	4.23 ± 0.90	3.35 ± 0.71	0.88 ± 0.20	3.82 ± 0.42	0.07 ± 0.02	0.19 ± 0.02
ScD_3-10	4.32 ± 0.35	3.43 ± 0.28	0.88 ± 0.17	3.96 ± 0.54	0.07 ± 0.01	0.20 ± 0.02
ScD_3-11	4.48 ± 0.38	3.49 ± 0.24	0.99 ± 0.17	3.45 ± 0.34	0.06 ± 0.02	0.13 ± 0.02
ScD_3-15	4.10 ± 0.85	2.90 ± 0.38	1.00 ± 0.20	2.94 ± 0.32	0.04 ± 0.01	0.18 ± 0.02
ScD_3-40	4.90 ± 0.70	3.74 ± 0.41	1.07 ± 0.27	3.74 ± 0.79	0.06 ± 0.02	0.14 ± 0.02
ScD_3-42	3.07 ± 0.44	2.33 ± 0.32	1.16 ± 0.29	3.13 ± 0.41	0.05 ± 0.01	0.14 ± 0.02
ScD_3-67	4.57 ± 0.52	3.68 ± 0.57	0.74 ± 0.13	4.92 ± 0.82	0.09 ± 0.02	0.18 ± 0.02
System I or electron transport mutants						
ScD_3-8	5.13 ± 0.62	3.97 ± 0.52	1.16 ± 0.26	3.42 ± 0.34	0.10 ± 0.02	0.24 ± 0.03
ScD_3-26	5.85 ± 0.41	4.52 ± 0.26	1.12 ± 0.11	4.04 ± 0.17	0.08 ± 0.01	0.28 ± 0.42
ScD_3-50	6.06 ± 0.35	5.02 ± 0.34	1.18 ± 0.06	4.20 ± 0.45	0.09 ± 0.02	0.14 ± 0.02

Photosystem II mutants produce a greater amount of the secondary carotenoids, such as hydroxyechininone, than the wild-type or the other mutant strains. It is to be stressed that the above analyses were made on dark-grown cells wherein the opportunity for photooxidation is minimal. The apparent increased oxidation status of the carotenoid fractions may indicate an alteration in chloroplast structure which is more accessible for degradatory oxidations in the Photosystem II mutants.

Total lipid analysis

Considerable importance has been given to the possible role of certain chloroplast lipids in the functioning of Photosystem II. Most critical attention has been given to those lipids which are abundant in the chloroplasts, the galactolipids, and particularly those that are enriched in the fatty acids, linolenic and linoleic acid[32,33]. Extensive effort has been spent on our part in attempts to determine if the Photosystem II mutants of Scenedesmus have an altered content of lipids; compounds studied included, in addition to the galactolipids, various phospholipids, sulfolipids, and neutral lipids. Repeated analyses with either ^{32}P- or ^{14}C-labeled algal lipid, or with non-radioactive fractions, have failed to show any major difference in the various lipid fractions. Both the amounts and the apparent rates of synthesis of the lipids in the wild-type and the mutants were similar. The various lipids separated from whole cells of Scenedesmus by the thin-layer chromatography procedure included the following: phosphatidyl inositol, sulfolipid, digalactosyl diglyceride, phosphatidyl glycerol, phosphatidyl ethanolamine, phosphatidic acid, monogalactosyl diglyceride and lecithin. No attempt was made to determine if any variation in fatty acid composition occurred in the various lipid fractions from either the wild-type or the various mutant strains.

Quinone analysis

Many researches have established that plastoquinone (most likely plastoquinone A (PQ-A)) functions in the electron transport system of photosynthesis. A more specific site of its function appears to be as one of the initial electron acceptors for Photosystem II. SMILLIE AND LEVINE[11] in studies on mutants of Chlamydomonas demonstrated that the plastoquinone content of two mutant strains, ac-115 and ac-141, (which have been shown to be Photosystem II mutants possessing many characteristics similar to those described here for the Photosystem II mutants of Scenedesmus) is decreased by about 80 %. Analyses on the plastoquinone content of the Scenedesmus mutants revealed a striking similarity in both mixotrophically and heterotrophically grown cells. All of the mutants, when grown heterotrophically, with the exception of ScD_3-42, contain approx. 50 % less plastoquinone than the wild type. When grown myxotrophically, the wild-type cells synthesize significantly more of the chloroplast pigments with a noticeable shift in the molar ratio of plastoquinone/chlorophyll. Only in mutants 40 and 42 was this ratio changed; the ratio in the other mutants remains essentially the same as that noted for heterotrophic growth. The data showing these results (Table IV) were obtained from 10 separate determinations on each of the algal strains. No significant differences were noted in the content of α-tocopherol or of α-tocopherol quinone. The low concentrations of vitamin K_1 made it extremely difficult to assess any variation between the wild-type and the mutant strains examined.

TABLE IV

MOLE RATIO PLASTOQUINONE A/CHLOROPHYLL IN HETEROTROPHICALLY AND MYXOTROPHICALLY GROWN CELLS OF SCENEDESMUS

Algal strain	*Heterotrophic*	*Mixotrophic*
Wild-type	0.0049 ± 0.0007	0.0093
System II mutants		
a′	0.0023 ± 0.0002	0.0027
4	0.0030 ± 0.0009	0.0024
5	0.0031 ± 0.0005	0.0028
10	0.0032 ± 0.0006	0.0029
11	0.0014 ± 0.0003	0.0028
15	0.0024 ± 0.0008	0.0025
40	0.0017 ± 0.0006	0.0044
42	0.0043 ± 0.0002	0.0056
67	0.0028 ± 0.0004	0.0038
System I or electron transport mutants		
8	0.0046 ± 0.0002	N.D.*
26	0.0029 ± 0.0001	N.D.
50	0.0044 ± 0.0002	N.D.

* N.D. = not determined.

Soluble protein components

The general properties of purified ferredoxin, ferredoxin–$NADP^+$ reductase, plastocyanin and cytochrome *c*-552 of *Scenedesmus obliquus* were examined and were found to be essentially identical to those obtained from other sources such as spinach chloroplasts[34], and Chlamydomonas[26]. No deficiency in any of the above listed factors was observed in the Photosystem II mutants. Following the isolation and purification of the individual factors, each of them was tested for its appropriate activity with chloroplasts isolated from wild-type cells[16] along with the appropriate cofactors also obtained from wild-type cells.

The data of Table V summarize our observations on these cofactors potentially involved in the reduction of $NADP^+$. It is clearly apparent that those derived from the Photosystem II mutants are fully active.

These findings substantiate the *in vivo* assays for photoreductive activity of the Photosystem II mutants where it was observed that a normal capacity for the reduction of carbon dioxide with hydrogen gas as the electron donor existed.

Membrane-bound electron transport components

The assay procedure employed for studying the water-soluble electron transport components of the chloroplasts of algae detects a number of other components which are probably not of chloroplastic origin. Cytochrome *c*-549, cytochrome *b*-562, and cytochrome *b*-558 are routinely detected during the purification of the other cofactors listed previously. The cytochrome *c* is undoubtedly of mitochondrial origin, but the source of the *b*-type cytochromes is questionable. Examination of the "lamellar float" obtained during the $(NH_4)_2SO_4$ fractionation of the soluble protein fraction reveals the electron transport components that remain bound to insoluble chloroplast mem-

TABLE V

THE ACTIVATION OF $NADP^+$ PHOTOREDUCTION OR CYTOCHROME PHOTOOXIDATION BY INDIVIDUAL ELECTRON TRANSPORT COMPONENTS OF CHLOROPLAST DERIVED FROM WILD TYPE OR MUTANT STRAINS OF SCENEDESMUS

The complete reaction mixture for the determination of NADP photoreduction contained wild type chloroplasts equivalent to 40 µg chlorophyll, 10 nmoles ferredoxin, 10 nmoles $NADP^+$ reductase, 5 nmoles plastocyanin, 6 µmoles $MgCl_2$, 50 µmoles Tris–HCl buffer (pH 7.2), and 0.8 nmole $NADP^+$, in 3 ml of 0.16 M sucrose–0.01 M KCl. Reduction rates are expressed as percent of control (wild type chloroplasts and cofactors) which was 228 µmoles $NADP^+$ reduced per h per mg chlorophyll. The reaction mixture for the measurement of the rate of photooxidation of cytochrome *c*-552 contained 5 nmoles reduced plastocyanin, 20 nmoles reduced cytochrome 552, 50 µmoles phosphate buffer (pH 7.0), 50 µmoles $(NH_4)_2SO_4$ and digitonized spinach chloroplasts equivalent to 10 µg of chlorophyll in a final volume of 3.0 ml. Oxidation rates are expressed as percent of control with cytochrome *c*-552 derived from wild-type Scenedesmus. Control rate was 72 µmoles cytochrome oxidized per h per mg chlorophyll.

Source of cofactor	*Limiting cofactor*			
	Ferredoxin	*$NADP^+$ reductase*	*Plastocyanin*	*Cytochrome f*
Wild-type	100	100	100	100
System II mutants				
ScD_3-a′	103	93	97	98
ScD_3-11	98	96	100	102
ScD_3-15	89	102	93	97
ScD_3-40	104	87	103	94
System I or electron transport mutants				
ScD_3-8	100	102	97	94
ScD_3-26	104	87	90	104
ScD_3-50	98	95	88	—

brane components. We have previously shown this fraction to contain bound cytochrome *f* (552 nm), and potentially two *b*-type cytochromes (559 and 562 nm)[25]. Only the cytochrome components contained in either acetone powders of the "lamellar float" or in acetone powders of the total algal extract (see METHODS AND MATERIALS) of mutants a′, 11, and 40 were examined. Slight amounts of chlorophyll retained in the acetone powders often obscured minor absorbance changes in the 520–530 nm region. However, as is shown in Fig. 1, the ferricyanide and dithionite difference spectra of the mutant strains compare favorably to those of the wild-type. The principle absorption maximum of the *b*-type cytochromes (as obtained in the dithionite difference spectrum) occurs at approx. 560 nm. With the techniques employed only this rather broad single peaked spectrum was obtained. Examination of the dithionite difference spectra in the Soret region showed only one peak at 431 nm in mutant and wild-type strains. To date, we have not seen a variation in either the bound or soluble forms of either the *b*-cytochromes or of cytochrome *f* comparable to that noted by SMILLIE AND LEVINE[11] for the Photosystem II mutants of Chlamydomonas. The difficulties involved in our assay procedures preclude any categorical statement about the possible interplay between plastoquinone A (membrane bound) and the *b*-type cytochromes which are thought to be involved in Photosystem II activity.

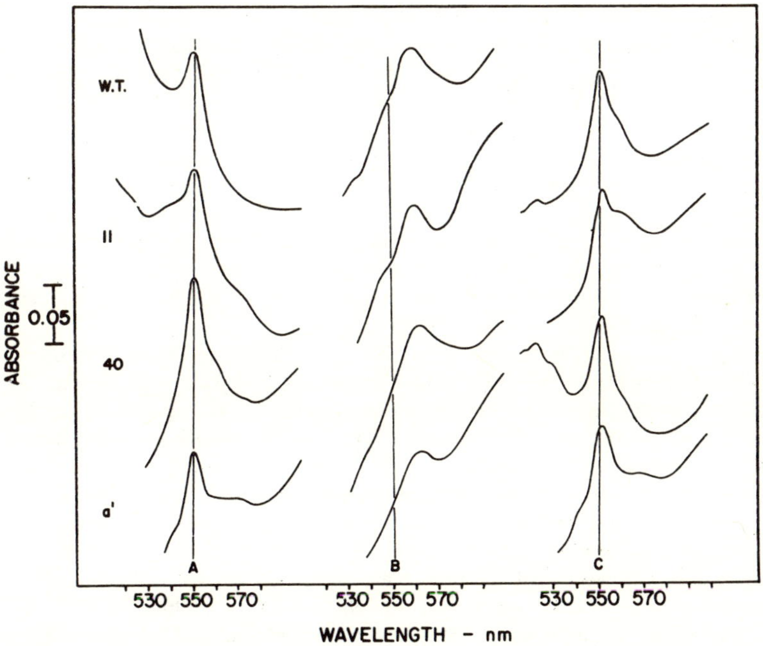

Fig. 1. Absorption difference spectra of lamellar protein extracts of wild-type, mutant 11, mutant 40, and mutant a'. 300 mg of acetone powder of insoluble lamellar fraction solubilized in 15 ml of 3% sodium dodecyl sulphate. Section A. Difference spectra of untreated sample *versus* ferricyanide treated. Section B. Difference spectra of dithionite treated sample *versus* normal sample. Section C. Difference spectra of dithionite *versus* ferricyanide treated sample. Solubilization of the acetone powder with 1% Triton X-100 (in 0.2 M Tris, pH 7.5) or 0.5% sodium deoxycholate did not alter the general features of the difference spectra obtained from that noted for the sodium dodecyl sulphate fractions.

Conclusions

As an approach to understanding the mechanism of action of Photosystem II, the application of the classic technique of biochemical genetics appeared to be a particularly useful approach when coupled with other experimental techniques employed in photosynthesis. In our studies on Scenedesmus, the methods employed for the selection of Photosystem II mutants have resulted in the isolation of nine mutants, some of whose characteristics are described in this paper. One of the more intriguing aspects of this study is the nearly identical behavior and characteristics observed in these strains. The features of lowered plastoquinone A, lowered total chlorophyll, and altered complementation of the carotenoids suggest that the change produced by the mutation, *i.e.* the loss of Photosystem II activity, is not necessarily the result of the loss of the ability of the alga to synthesize a single factor, such as plastoquinone, but perhaps rather to the loss of a discrete part of the Photosystem II unit within the chloroplast. The results of the *in vivo* and *in vitro* analyses for the activity of the separate photosystems (Table I and II) show clearly that the Photosystem II mutants lack any demonstrable partial reaction related to the reductive portion of this photosystem. The inability of chloroplasts of these mutants to photoreduce DCIP with an exogenous electron donor (diphenylcarbazide) correlates with our observations on the lowered plastoquinone content of these mutants. The absence of PQ-A in Photosystem II should cause the loss of the reductive activity of this system. Since other electron donor systems, such as the DCIP–ascorbate couple, can by-pass the plastoquinone site in the photoreduction $NADP^+$, isolated chloroplasts of the Photosystem II mutants should have the capacity for this reaction as was observed.

The ability of whole cells of the Photosystem II mutants to photoreduce carbon dioxide similarly indicates the non-essentiality of this portion of Photosystem II in photoreduction, a conclusion which is also apparent from the lack of sensitivity of photoreduction to DCMU.

Evidence has been obtained by a number of investigators that plastoquinone may also function in Photosystem I or in cyclic photophosphorylation[13]. The finding that plastoquinone synthesis is not completely blocked in the various mutants perhaps reveals that some portion of the plastoquinone pool is functional in Photosystem I. However, the conclusion that plastoquinone functions in two sites is difficult to comprehend since in our analysis only plastoquinone A was found (in healthy young cells) and the separation of this pool into two separate physical pools, each differing grossly in physicochemical properties, does not appear feasible at this time. If plastoquinone functions at only one site, then the presence of limited amounts of PQ-A may be representative of either that contained in nonphotosynthetic inclusions of the cell or of the photosynthetically incompetent portions of Photosystem II. More definitive studies with various mutant phenotypes are in progress in order to test the hypothesis of multiple sites of action of the plastoquinones.

Additional evidence gained from fluorescence studies will be presented in a subsequent paper which supports the interpretation that the Photosystem II mutants lack the reducing portion of this photosystem. A more comprehensive interpretation of the results of this paper will be presented after documentation of the fluorescence properties of the Photosystem II mutants has been presented.

ACKNOWLEDGMENTS

Support for the research reported here was provided in part by grants from the U. S. Public Health Service (GM-11745) and the Atomic Energy Commission (AT(454)-1783). The authors wish to thank Mrs. Marjorie Sherrod, Mrs. A. J. Powells, and Miss Janet Williams for the technical assistance throughout the course of this study.

REFERENCES

1 S. Granick, *Ann. Rev. Plant Physiol.*, 2 (1951) 115.
2 H. Claes, *Z. Naturforsch.*, 13b (1958) 222.
3 S. Bendix and M. B. Allen, *Arkiv Mikrobiol.*, 41 (1962) 115.
4 D. von Wettstein, *The Photochemical Apparatus, Its Structure and Function, Brookhaven Symp. Biol.*, 11 (1958) 138.
5 R. Sager, *The Photochemical Apparatus, Its Structure and Function, Brookhaven Symp. in Biol.*, 11 (1958) 101.
6 L. Bogorad, F. V. Mercer and R. Mullens, *Photosynthetic Mechanisms in Green Plants, Natl. Acad. Sci. Natl. Res. Council Publ.*, 1145 (1963) 560.

7 R. Sager and M. Zalokar, *Nature*, 182 (1958) 98.
8 S. Volk and N. I. Bishop, *Photochem. Photobiol.*, 8 (1968) 213.
9 N. K. Boardman and H. R. Highkin, *Biochim. Biophys. Acta*, 126 (1966) 189.
10 I. C. Anderson and D. S. Robertson, *Plant Physiol.*, 35 (1960) 531.
11 R. M. Smillie and R. P. Levine, *J. Biol. Chem.*, 238 (1963) 4058.
12 J. Lavorel and R. P. Levine, *Plant Physiol.*, 43 (1968) 1049.
13 R. P. Levine, *Ann. Rev. Plant Physiol.*, 20 (1969) 523.
14 N. I. Bishop, *Record Chem. Progr.*, 25 (1964) 181.
15 N. I. Bishop, *Nature*, 195 (1962) 55.
16 L. H. Pratt and N. I. Bishop, *Biochim. Biophys. Acta*, 153 (1968) 664.
17 E. C. Weaver and N. I. Bishop, *Science*, 140 (1963) 1095.
18 E. C. Weaver and N. I. Bishop, *Photosynthetic Mechanisms in Green Plants, Natl. Acad. Sci. Natl. Res. Council Publ.*, 1145 (1963) 138.
19 W. Bertsch, J. R. Azzi and J. B. Davidson, *Biochim. Biophys. Acta*, 142 (1967) 129.
20 W. L. Butler and N. I. Bishop, *Photosynthetic Mechanisms in Green Plants, Natl. Acad. Sci. Natl. Res. Council Publ.*, 1145 (1963) 91.
21 H. Senger and N. I. Bishop, *Plant Cell Physiol.*, 7 (1966) 441.
22 M. Holden, in T. W. Goodwin, *Chemistry and Biochemistry of Plant Pigments*, Academic Press, New York, 1965, p. 466.
23 G. MacKinney, *J. Biol. Chem.*, 140 (1941) 314.
24 F. L. Crane, *Ciba Found. Symp. Quinones in Electron Transport*, 1961, p. 68.
25 R. Powls, J. Wong and N. I. Bishop, *Biochim. Biophys. Acta*, 180 (1969) 490.
26 D. S. Gorman and R. P. Levine, *Plant Physiol.*, 41 (1966) 1637.
27 D. S. Gorman and R. P. Levine, *Plant Physiol.*, 41 (1966) 1643.
28 H. Matsubara, *J. Biol. Chem.*, 243 (1968) 370.
29 T. Yamashita and W. L. Butler, *Plant Physiol.*, 43 (1968) 1978.
30 L. P. Vernon and E. R. Shaw, *Biochem. Biophys. Res. Commun.*, 36 (1969) 878.
31 R. H. Nieman, H. Nakamura and B. Vennesland, *Plant Physiol.*, 34 (1959) 262.
32 J. Erwin and K. Bloch, *Biochem. Biophys. Res. Commun.*, 9 (1962) 103.
33 J. Erwin and K. Bloch, *Biochem. Z.*, 338 (1963) 496.
34 S. Katoh, I. Shiratori and A. Takamiya, *J. Biochem. Tokyo*, 51 (1962) 32.

THE INFLUENCE OF TWO NATURALLY OCCURRING FACTORS, FROM THE LEAVES OF *RICINUS COMMUNIS*, ON THE FLUORESCENCE INDUCTION OF HIGHER PLANT CHLOROPLASTS AND INTACT ALGAE

MARCIA BRODY

ABSTRACT Changes in fluorescence induction, brought about by incubation of chloroplasts (*Zea mays*) in an aqueous extract of *Ricinus* leaf, have been divided, on the basis of speed of manifestation, into two categories: "fast" changes and "slow" changes (i.e. those observed after 5 min and 1½ hr of incubation, respectively). The former, which include a large increase in the magnitude of the fast component of variable fluorescence and a retardation of decay from maximum to minimum levels of fluorescence, have been ascribed to inhibition of electron transport at a site beyond that of 3-(*p*-chlorophenyl)-1,1-dimethylurea (CMU)—i.e., towards system I; these changes result from the action of a fraction of the extract consisting of molecules of small size. The latter changes, which include a marked attenuation of the variable part of fluorescence induction, have been associated with system II and may arise from inhibition of electron flow between water and Q or from decrease in number of functional reaction centers; these changes result from the activity of a proteinaceous fraction of the extract, that simultaneously converts the low temperature steady-state emission spectrum of the chloroplasts into a one-banded one, with maximum at 698 nm.

INTRODUCTION

Because the 698 nm, low temperature fluorescence band (1–7) is thought to originate from a form of chlorophyll active in photosystem II, perhaps as the energy trap (3–6, 8), it has generated considerable interest. Recently, Brody, Nathanson, and Cohen (9) reported changes in spectra of steady-state fluorescence of chloroplasts (isolated from various organisms) incubated in aqueous leaf extract of castor oil plant (*Ricinus communis*). As revealed by measurements at 77°K, the three-banded emission spectrum, with maxima at about 685, 698, and 735 (or 720) nm, was converted, as a function of time and temperature of incubation, into an essentially one-banded system, with maximum at about 698 nm. In the present work is reported the influence of this extract on the induction of fluorescence in isolated chloroplasts of higher plants and in intact algae. This influence appears to be of dual nature, and may involve two separate components in the extract, acting at different sites and at different rates.

The first action of the extract is very rapid, and results in the following changes in fluorescence induction: (*a*) a small increase in maximum fluorescence yield, primarily in the variable portion, (*b*) a large increase in amplitude of the fast part of the rise of variable fluorescence, leading to more rapid achievement of maximum yield, and (*c*) a progressive decrease in the rate of decay from the fluorescence peak to the stationary level, accompanied by a rise in the stationary level itself.

The second action of the extract is manifested more slowly, and appears to be related to the similarly time-dependent changes in steady-state fluorescence referred to above (9); it results in the following changes in fluorescence induction: (*a*) a marked decline in fluorescence yield, primarily in the variable portion, (*b*) a retardation of the slow part of the rise in variable fluorescence, and (*c*) some loss of the initial increase in amplitude of the fast part of variable fluorescence.

Since preliminary experiments revealed that some of the changes in fluorescence induction brought about by the extract were similar to some produced by CMU, the action of this herbicide was also studied, for purposes of comparison.

In the discussion section of this paper, interpretations of the data are given in terms of possible sites of action of the extract in the electron transport chain of photosynthesis. It is suggested that the faster acting effect of the extract blocks electron transport at a site beyond that where CMU blocks, i.e. towards system I (perhaps *within* system I, per se). It is also suggested that the second or slower effect of the extract on fluorescence induction results from an inhibition of electron transport somewhere between water and Q, or from a decrease in number of functional system II reaction centers.

Preliminary experiments on the molecular size (and nature) of the active components of the extract are described. Of the two types of molecules resulting from Sephadex G-25 separation, the smaller appears to be associated with the fast changes in fluorescence induction, while the larger appears to be associated with the slow changes (and also with the transformation of the steady-state emission spectrum).

MATERIALS AND METHODS

Aqueous extracts of leaves of *Ricinus communis* were prepared as described earlier (9) (by a modified method of McCarty and Jagendorf [10]) from greenhouse-grown plants of variety Baker 296 (seeds courtesy of Dr. R. L. Ory, Seed Protein Laboratory, Department of Agriculture, New Orleans, La.) or variety zanzibariensis (seeds purchased from Burnett Brothers, New York). The medium used for both extraction of leaves and isolation of chloroplasts (to be referred to as E-I medium) was composed of 0.35 M NaCl and 0.04 M Trizma-HCl (source of Trizma: Sigma Chemical Co., St. Louis, Mo.) pH 7.8 or 7.2. Leaves of greenhouse-grown dwarf corn (*Zea mays*, hybrid variety INRA 200) were used as the source of chloroplasts (see references 9 and 11 for isolation procedure). Intact *Chorella* cells were centrifuged out of their culture medium (12) and resuspended in a small volume of E-I medium. At the beginning of each experiment, predetermined quantities (less than 2 ml) of chloroplasts or algae, at 5°C, were mixed (in the reservoir of the fluorescence induction apparatus) with 250 ml of suspending medium, also at 5°C. This suspending medium was either E-I medium, or *Ricinus* leaf extract, the latter henceforth to be abbreviated RLE. In any one series of experiments, separate and equal aliquots of the same plant materials were used for treatment and control. Suspensions of algae or chloroplasts were prepared so as to yield final chlorophyll concentrations of 10–50 μg chlorophyll ($a + b$) per ml (as determined by the method of Arnon [13]).

Although fluorescence induction measurements were begun immediately after mixing, the data presented here are for suspensions which had reached 25°C (see footnote 2); this warming-up to constant temperature took less than 5 min. The stop-flow apparatus used for the fluorescence induction studies has been described previously (14). Excitation was with a band of blue light isolated from a high pressure 900 w xenon arc lamp by means of a transmission grating monochromator set at 480 nm. A half-band width of 10 nm was utilized. In the experiments reported here, the light was attenuated by a neutral density filter of metalized glass, which transmits 44.8% of the available light; under these conditions, the exciting intensity was about 9×10^3 ergs/cm^2 per sec, and may be considered to be "moderate" compared to the designation "low" or "high" used by Lavorel (14). In the results section of this paper, the data given in the figures are taken directly from recorder tracings, therefore the values of O fluorescence levels (see below) are apparent ones;[1] actual O levels are given in the corresponding tables. (Corrections were based on measurements of fluorescence as a function of the intensity of the exciting light.) The wavelength at which fluorescence induction was monitored was 685 nm; a red cut-off filter (Corning glass 2-58, Corning Glass Works, Corning, N. Y.), complementary to the exciting light, was placed before the analyzing Bausch & Lomb monochromator (Bausch & Lomb Optical Products, Bausch & Lomb, Inc., Rochester, N. Y.), the slits of which were set to pass light of half-band width of $\simeq$10 nm. The speed of recorder chart paper was generally 120 mm/min; however, rise and decay times were also recorded at speeds of 300 and 30 mm/min, respectively.

For experiments involving CMU, aliquots of 0.1 ml of stock solution (1×10^{-2} M in absolute ethanol) were delivered to the 250 ml suspension in the reservoir, to yield concentrations up to 8×10^{-5} M CMU.

During the course of the fluorescence induction measurements, small samples (about 2 ml) were periodically removed from the reservoir of the apparatus, so that concurrent determinations of steady-state emission could be made. Excitation of steady-state fluorescence was accomplished by an optical system which passed blue light (the 436 nm line) isolated from a 100 w mercury arc lamp by Corning glass filters 3-73 and 5-58 (Corning Glass Works). The sample vessel was lowered into a flat bottom Dewar flask filled with liquid nitrogen, and fluorescence was both excited and detected from the same (bottom) surface of the sample. Fluorescence spectra were measured with the aid of a Bausch & Lomb monochromator (Bausch & Lomb Inc.) (half-band width 6.6 nm) in combination with a red cut-off filter, Corning glass 2-58. A Dumont 6911 multiplier phototube (Dumont Laboratories, Fairchild Camera and Instrument Corp., Clifton, N. J.) was used as detector, and the resulting signal was measured with a sensitive galvanometer.

---

[1] In the range of exciting light intensity used, the level in O state is invariably overestimated with the flow method (14), the more so as the OP photochemical process (see below) increases in activity; see also J. Lavorel (15), where this relationship is considered in the case of *Chlorella* treated with CMU.

The Sephadex G-25 (coarse grade), used for chromatography of crude RLE was obtained from Pharmacia Fine Chemicals Inc., Uppsala, Sweden. Before introduction into the column, the Sephadex was washed and swollen, and then equilibrated with E-I medium.

As an aid to understanding the results presented below, there are given in Figs. 1 *A* and *B* schematic diagrams of the types of fluorescence induction curves obtained with control chloroplasts and control algal cells, respectively. The various parameters to which reference will be made are labeled either in accordance with earlier usage (16), or are newly designated. Their meaning is as follows: ***d*** indicates zero fluorescence (dark level); *c*, in relative fluorescence units (mm intervals on the chart paper) represents the constant portion of fluorescence, obtained when the suspension is flowing past the exciting beam; *v*, obtained when the suspension is stopped in the light beam, represents the contribution of the variable portion to total fluorescence, $c + v$; *v* is composed of a fast part, beginning at the *m* or *O* level and extending to the break or inflection point, ***I***, and a slow part, beginning at *I* and rising to the ***P*** level, which represents maximum or peak fluorescence; 25%, 75%, and 100% rise time represent the time (in seconds) for these respective percentages of the *OP* rise to be realized; in chloroplasts, 10% "*PO*" decay time is used here to refer to the time required for the variable portion of fluorescence to fall 10% of its value from the *P* level (measured from the time at which ***P*** is initially attained) to the previously determined *O* level (the reader should note that this time interval does *not* correspond to that required for 10% of the *PO* transition, which occurs when the suspension is allowed to flow again); in intact algae, *S* represents that stationary state to which the yield drops in the stopped system; the *S* level is monitored by the magnitude of the *PS* transition; 20%, 40%, and 100% *PS* decay times refer to the intervals required to achieve these percentages of the *PS* decay. The following example is given to demonstrate the manner in which rates of transition (Tables I and II) are calculated: the rate of the 75% ***OP*** rise is obtained by dividing the time it takes to achieve

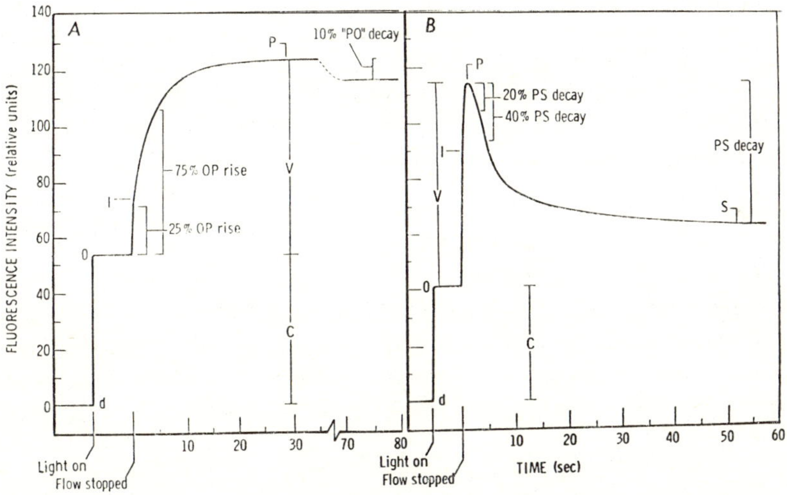

FIGURE 1 Schematic diagrams of fluorescence induction of *Zea mays* chloroplasts (A) and intact cells of *Chorella* (B). Zero fluorescence is given by *d* (dark level); for definition of labeled parameters, refer to body of paper (section on Materials and Methods). Note break in continuity of abscissa (time scale) in Fig. 1 A.

75% of the *OP* rise by the magnitude of this portion of the rise (i.e. 75% *OP* = 75% *v*) in relative fluorescence units (mm); rates are given, therefore, in terms of sec/fluorescence unit. One would like to measure, with the stopped suspension of chloroplasts, the *PS* transition (the "*PO*" measurement being a rather less desirable substitute) but such a measurement is practical only in the case of intact algae; it is impractical in the case of isolated chloroplasts. in which the *PS* decay is extremely slow.

RESULTS

A Comparison of the Influence of RLE and CMU on Steady-State Fluorescence and Fluorescence Induction in Chloroplasts

The Influence of RLE. Data of the type presented in Table I were obtained in the following way. Chloroplasts prepared on any one day were divided into two aliquots. Fluorescence induction was monitored for the control aliquot (chloroplasts in E-I medium) during a period of 1½ hr. Since very little change with time (generally a decrease of ≈5%) was observed in the control, data are given only for the measurement made 5 min after mixing.[2] Fluorescence induction was also monitored for the second aliquot (chloroplasts in RLE) during a period of 1½ hr; in this case, data are given both for measurements made 5 min and 1½ hr after mixing.

Although the data shown here are representative of the kinds of changes brought about by RLE, the degree of change was found to vary somewhat from day to day, and will be commented upon where appropriate. While a part of these variations reflect, no doubt, dissimilarities in the chloroplast preparations per se (i.e. in the controls), the remainder may be attributed to differences in concentration of active material in the RLE's which were freshly prepared each day. To facilitate comparison, the data have additionally been expressed in terms of ratios (experimental-to-control).

The reader may wish to refer to Fig. 2, in conjunction with Table I, for in this figure may be found recorder tracings of these fluorescence induction curves.

Although the data given in the present paper are for plant materials suspended in E-I media (or RLE) of pH 7.8, essentially similiar results were found at pH 7.2.

Concurrent steady-state fluorescence emission measurements, made at 77°K, revealed that in the case of chloroplasts suspended in RLE, the 685 nm band was disappearing as a discrete entity (or had largely "disappeared"), 5 min after mixing, while at 1½ hr after mixing, the 698 nm band was the major one which could be detected. No such changes in steady-state fluorescence emission were observed in the controls.

We will now turn our attention back to Table I (and Fig. 2), to consider, in

[2] Temperature, in common with several other factors (see reference 15) has a strong influence on the relative rates of the light-dependent and light-independent reactions associated with fluorescence induction; at 5°C, almost all of the rise in variable fluorescence is of the fast type, i.e., there is a similarity of effects of low temperature and RLE on the fast component of variable fluorescence.

TABLE I

A COMPARISON OF THE EFFECTS OF RLE AND CMU TREATMENT ON FLUORESCENCE INDUCTION OF *ZEA MAYS* CHLOROPLASTS

Fluorescence induction parameters	Control C	Period of incubation in RLE				CMU (added to control) 8×10^{-5} M	
		5 min		1½ hr			
		Experimental E	$\frac{E}{C}$	Experimental E	$\frac{E}{C}$	Experimental E	$\frac{E}{C}$
1. v (mm)	70.0	86.0	1.23	41.0	0.585	52.4	0.748
2. c (mm)	61.5	69.0	1.12	56.0	0.911	64.6	1.05
3. $c + v$ (mm)	131.5	155.0	1.18	97.0	0.737	117.0	0.89
4. OI (mm)	14.0	60.0	4.28	16.0	1.14	45.9	3.28
5. OI/v	20.0%	69.8%	3.49	39.0%	1.95	87.7%	4.39
6. $OI/c + v$	10.6%	38.7%	3.72	16.5%	1.56	39.2%	3.70
OP rise							
7. 25% *OP*							
a*	0.500	0.250	0.500	0.500	1.00	0.150	0.300
b*	0.029	0.012	0.400	0.049	1.69	0.012	0.397
8. 75% *OP*							
a	7.50	1.25	0.167	2.50	0.333	0.500	0.067
b	0.143	0.019	0.136	0.081	0.563	0.013	0.089
9. 100% *OP*							
a	25.0	7.50	0.300	18.0	0.72	2.00	0.080
b	0.357	0.087	0.244	0.438	1.23	0.038	0.107
"*PO*" decay							
10. 10% "*PO*"							
a	45.0	56.0	1.24	85.0	1.89	90.0	2.00
b	6.43	6.50	1.01	20.7	3.22	17.2	2.68

* a, time (sec); b, rate (sec/mm).

relation to the controls, the "fast" and "slow" changes in fluorescence induction brought about by RLE. These represent, respectively, changes which are detected 5 min and 1½ hr after mixing.

In respect to total fluorescence, the following (see Table I, item 3) may be observed: there is a small increase in magnitude of $c + v$ when chloroplasts are first mixed with RLE; however, after 1½ hr this maximum yield has fallen to about three-quarters of the control value. From items 1 and 2, it may be observed that the initial increase in $c + v$ results, in greater part, from increase in variable fluorescence; the subsequent fall in total fluorescence may also be predominantly attributed to attenuation of the variable portion. In some preparations, v was found to fall to about 40% of its control value after 1½ hr in RLE.

A very striking change, seen upon mixture of chloroplasts and RLE, is marked enhancement of the amplitude of the fast part, *OI*, of the rise in variable fluorescence (see items 4–6). (In the control, *OI* generally constitutes about ⅕ to ¼ of v.) This increase in magnitude (usually two- to fourfold) of the *OI* phase results in accelerated achievement of the *OP* transition, which may be seen at both the 75%

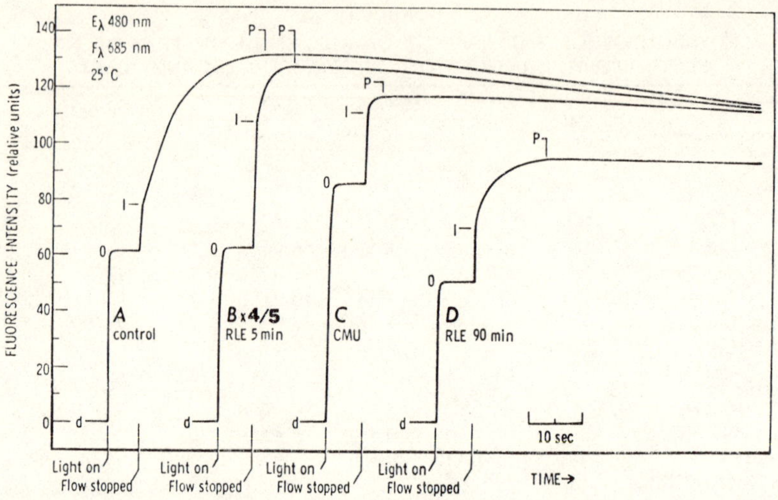

FIGURE 2 Fluorescence induction of *Zea mays* chloroplasts. Curve A, control; curves B and D, chloroplasts incubated in RLE for 5 min and 1½ hr, respectively; curve C, CMU-treated chloroplasts (8×10^{-5} M). Curves A–D are tracings of recordings; curve B has been multiplied by 4/5 for ease of comparison with other curves. Fluorescence excited at 480 nm and monitored at 685 nm. Recorder chart speed was 120 mm/min; abscissa time units indicated by scale of 10 sec.

and peak levels (items 8 and 9); an additional factor in the accelerated rise to maximum fluorescence is the increase in slope of the fast component, reflected in the reduction in time to achieve the 25% level of the *OP* rise (item 7). With increasing time in RLE, the contribution of the *OI* phase diminishes, but generally remains higher than in the control (see items 4–6). Note the influence of prolonged exposure to RLE on 100% *OP* rise times (items 7–9); although the 75% *OP* rise time remains short (because of the increased amplitude of the *OI* transition) the *IP* rise gets slower and slower. It was noted in some preparations that because of this phenomenon, the 100% *OP* rise time was considerably longer—and the rate of rise considerably slower (sometimes two to three times slower)—in chloroplasts exposed to RLE for extended periods of time than in control chloroplasts.

The already slow decay from the fluorescence peak *P*, in control chloroplasts is further arrested by suspension of the chloroplasts in RLE; the time required to reach the 10% "*PO*" decay level increases with time in extract (see item 10). This retardation by RLE can be observed more readily in the case of intact *Chlorella* cells (to be described in detail, below), in which the decay to the actual *S* level is very rapid in the control situation. (See Fig. 3 and Table II.)

The Influence of CMU. Aliquots of CMU were added to the control, subsequent to the fluorescence induction measurement made 1½ hr after mixing (see Table I); these data have also been expressed in terms of ratios (experimental-to-control).

Low temperature, steady-state fluorescence emission spectra were determined with such CMU-treated chloroplasts and it was found that, even after equally long periods of incubation in CMU, the types of spectral changes which accompany RLE incubation are not observed.

The influence of CMU on fluorescence induction of chloroplasts will now be compared with that of RLE (see Table I and Fig. 2). The presence of 4×10^{-6} M CMU has little effect on total fluorescence. However, with increasing concentration, the apparent level of *O* increases, and the sum of $c + v$ decreases. At 8×10^{-5} M CMU total fluorescence has fallen to about 90% of the control (see Table I, item 3, and Fig. 2), the change here, too, resulting primarily from decrease in variable fluorescence (see items 1 and 2).

In 8×10^{-5} M CMU almost all of the *OP* rise is composed of the fast component *OI* (items 4–6), the slow rise having been practically obliterated. (The magnitude of *OI* is only slightly smaller in 4×10^{-6} M CMU than in 8×10^{-5} M CMU.)

As a consequence of the increase in magnitude of *OI* (and perhaps also increase in its slope—see item 7) the *OP* rise times are shortened (items 7 *a*, 8 *a*, and 9 *a*) and the rates of rise accelerated (see items 7 *b*, 8 *b*, and 9 *b*) as they are with chloroplasts exposed to RLE for 5 min. That the rates of rise in CMU are faster than in RLE (especially to 100% *OP*) is due largely to the contribution of the slow component (*IP*) of variable fluorescence in the latter. Indeed the persistence of the *IP* transition and its marked deceleration with increasing time in RLE constitute two of the major differences in the effects of RLE and CMU. Also, in the case of CMU addition, the time required to reach the 10% "*PO*" decay level is increased, just as it is in the case of chloroplasts suspended in RLE (see item 10).

The Influence of RLE on Steady-State Fluorescence and Fluorescence Induction in Chlorella

Refer to Fig. 3, in which are presented tracings of fluorescence induction curves for *Chlorella* suspended in E-I medium (control), or in RLE for periods of 15 and 45 min; the salient features of these curves have been summarized in Table II. The level of the inflection point, *I*, and, therefore, the exact magnitude of the fast component, *OI*, of the *OP* rise is difficult to determine. In such cells, the rise in variable fluorescence is very fast and there is present little slow component (compare with Fig. 2); however, it may be that here, too, RLE is increasing the amplitude of the fast component (item 4 in Table II) and the proportion which this fast component contributes to variable fluorescence (see items 5 and 6). The situation in regard to the *OP* rise is even less clear (item 7). The most pronounced change brought about by RLE in *Chlorella* cells is in the decay of fluorescence from the peak to the stationary level. Although the actual time it takes to reach the *S* level decreases in the

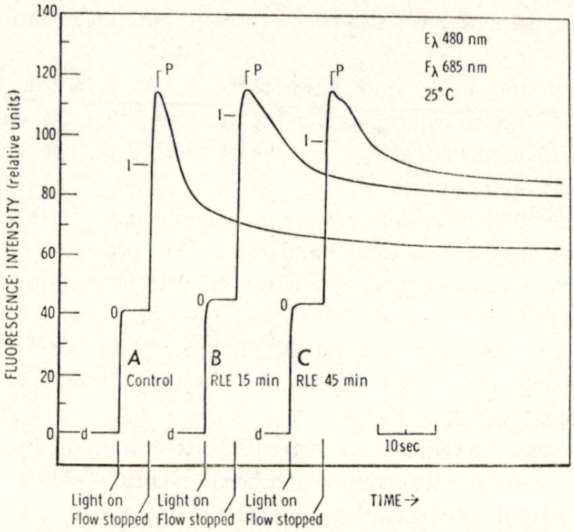

FIGURE 3 Tracings of fluorescence induction recordings of intact *Chlorella* cells. Curve A, control; curves B and C, cells incubated in RLE for 15 and 45 min, respectively. Other conditions as in Fig. 2.

presence of RLE (see item 13 *a*), the level of *S* itself has risen towards *P* (see items 8, 9 and 10). Thus, the rate of decay has actually slowed down (see item 13 *b*). This phenomenon may be additionally observed in the time it takes to decay 20 % and 40 % from the peak to the *S* level (items 11 *a* and 12 *a*), as well as in the respective rates of decay (items 11 *b* and 12 *b*).

The major difference in the effect of RLE on fluorescence induction in whole *Chlorella* cells and on corn chloroplasts is that in the former, the "slow" changes in fluorescence induction do not seem to occur to any appreciable extent. For example, in intact cells the marked decrease in fluorescence yield, especially in the variable portion, is not observed (see items 1–3).

Steady-state emission spectra (77°K) were determined for intact *Chlorella,* both for the control case, and for RLE-treated cases. It was found that, even after 45 min in RLE at 25°C, few spectral changes had occurred. Although the 685 nm band had "disappeared" the ratio of fluorescence intensities, F 725/F 698, had decreased to only 1.7 in comparison to a ratio of 1.9 in control cells. (It has subsequently been determined with chloroplasts of *Chlorella* that this ratio decreases to 0.6 after 45 min of incubation in RLE at 25°C.)

The Size and Nature of the Active Components in RLE

In a preliminary attempt to determine the nature of the active components (i.e., those capable of giving rise to the observed fluorescence induction changes), crude

TABLE II

THE EFFECT OF RLE ON FLUORESCENCE INDUCTION OF *CHLORELLA* CELLS

Fluorescence induction parameters	Control C	Period of incubation in RLE: 15 min, Experimental E	15 min, $\frac{E}{C}$	45 min, Experimental E	45 min, $\frac{E}{C}$
1. v (mm)	73.0	70.0	0.96	71.0	0.973
2. c (mm)	42.0	45.8	1.09	43.5	1.03
3. $c + v$ (mm)	115.0	115.8	1.01	114.5	0.996
4. OI (mm)	49.0	59.0	1.21	55.0	1.12
5. OI/v	67.0%	84.0%	1.25	77.5%	1.16
6. $OI/c + v$	42.6%	51.0%	1.20	48.0%	1.13
7. 100% OP rise:					
a*	1.50	1.80	1.20	2.00	1.33
b*	0.21	0.026	1.250	0.028	1.375
8. PS	52.5	34.0	0.648	30.0	0.57
9. PS/v	71.8%	48.5%	0.675	42.3%	0.59
10. $PS/c + v$	45.6%	29.4%	0.645	26.2%	0.575
PS decay					
11. 20% decay					
a	1.80	2.80	1.56	2.80	1.56
b	0.171	0.412	2.40	0.467	2.73
12. 40% decay					
a	3.00	5.00	1.67	4.80	1.60
b	0.143	0.368	2.57	0.400	2.80
13. 100% decay					
a	50.00	34.00	0.681	43.00	0.86
b	0.954	1.00	1.05	1.43	1.50

* a, time (sec); b, rate (sec/mm).

RLE was placed in a cellulose sac and dialyzed against a total of 60 times its own volume of E-I medium (six changes of medium during a 24-hr period) in the cold (5°C). The contents of the sac (predominantly larger molecules), when tested on chloroplasts, gave only very weakly the "fast" changes, but much more strongly the "slow" changes. A second dialysis experiment was made, in which the crude extract, in the sac, was dialyzed against three times its volume of E-I medium for a period of 18 hr. When the nondialyzed portion was diluted to a volume equal to that of the dialyzate, and the two aliquots tested, they gave about equal activity for fluorescence induction changes of the "fast" kind, but changes of the "slow" kind were again only given by the nondialyzed portion.

When crude RLE was chromatographed on a Sephadex G-25 column, two distinctly different fractions were collected; these were tested on chloroplasts for fluorescence induction activity of the "fast" type. The first fraction, consisting of large molecules, came quickly through the column. It was colorless and turbid and yielded a broad absorption band (between 260 and 280 nm) in the near ultraviolet.

This fraction gave only slight activity of the "fast" kind; however, it did bring about changes in steady-state emission similar to those produced by the crude extract. The second fraction, consisting of smaller molecules, was removed only very slowly from the column. It was clear, pale lemon-yellow in color, and slightly fluorescent (orange-yellow). It yielded an absorption spectrum consisting of a narrow band with maximum at 270 nm and a secondary (lower and broader) band at 360–380 nm with a long "tail" out to about 440 nm. This fraction, when tested on chloroplasts, showed strong fluorescence induction activity of the "fast" type; low temperature steady-state emission characteristics were not modified in its presence.

DISCUSSION

In the results section of the paper it was seen that two types of changes in fluorescence induction are brought about by RLE. Even though these are probably occurring simultaneously at different rates, "fast" and "slow" changes will be discussed separately.

A frequently used model will be utilized to interpret the data given above:

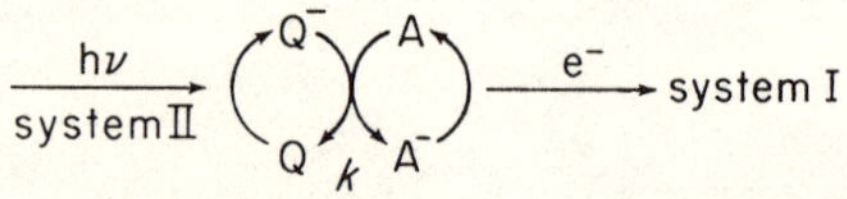

In the model, Q may be thought of as the quencher of Duysens and Sweers (17), or alternately as Joliot's E (see references 16–18); when Q (the primary oxidant or electron acceptor of system II) is in the Q^- state it cannot quench the excitation of chlorophyll$_{II}$ and therefore, the reduction (by the strong primary reductant) of Q to Q^-, brought about by events following absorption of light by system II, results in the rise of variable fluorescence from the *O* to the *P* level; refer to Fig. 1. The fast component of this rise, *OI*, represents the direct photochemical conversion (19) of Q to Q^- (see, however, references 20 and 21 for the involvement of activation, as well as reduction, in this process), while the slow component of this rise, *IP*, represents the slowing down of the Q to Q^- conversion because of continued restoration of Q by the light-independent A to A^- reduction (A being the large pool of secondary oxidant); hence, this second part of the rise is largely nonphotochemical. The reoxidation of A^- to A (present at about 20 times the concentration of Q) presumably results from electron transfer to system I, to which it is coupled by a chain of intermediates. The decay of fluorescence from the *P* to the *S* level is probably associated with the formation of a quencher distinct from Q; see, for example, the suggestion of Duysens and Sweers (17). (Munday and Govindjee (22) have alternatively suggested that this decline could be associated with induction of the Calvin cycle.)

Inhibition of system II by CMU is accompanied by augmentation of the *OI* transition, suppression of the *IP* transition, and retardation of the decay from the *P* level—phenomena which have been interpreted (17, 23) in terms of the herbicide acting between Q^- and A (the break in electron flow to system I presumably keeping fluorescence at its maximal value *P*). Seen among the first effects of RLE is also a more rapid reduction of Q. However, RLE does not suppress the *IP* rise, while CMU does. It is not certain, at the present time, if the fast acting component of RLE behaves as a nonsaturating concentration of CMU—at which the *IP* phase remains always visible but smaller, or if it (more likely) acts in some way or at some site different than CMU, so that *IP* is not suppressed. Decrease in the rate of reaction between Q^- and A, or partial inactivation of A, would result in less slowing of the Q to Q^- conversion (and, therefore, in increased *OI* transition).

Also observed among the first effects of RLE on fluorescence induction is a (small) increase in maximum fluorescence yield, principally in the variable portion. This may reflect a larger concentration of Q^-, perhaps as a result of conversion of some of the quencher distinct from Q, into Q itself, and/or as a result of a "backup" of electrons from a break in the electron transport chain to system I. In regard to the latter, it has been noted in experiments on light-induced changes in absorption of chloroplasts suspended in RLE (24), that there is present a component which affects system I (enhancing the changes at low concentrations and inhibiting them at higher concentrations). Furthermore, studies of oxygen transients (25) reveal an immediate decrease in the photochemical activity of system I upon introduction of RLE to suspensions of chloroplasts.

The slowing up of the fluorescence decay, *PS* or "PO," which results in the yield being kept close to its maximum level, may be interpreted in terms of the disruption of electron transport from system II to system I, as above, or, on the basis of the rise towards *P* of the *S* level itself, especially evident in the case of *Chlorella* (see also reference 26), a diminution in the amount of quencher distinct from Q.

In their 1968 paper on the effect of manganese deficiency on fluorescence properties of chloroplasts, Anderson and Thorne (27) reported an enhancement in the 77°K ratio F 735/F 693 (6.2 in the deficient as compared with 2.3 in the control case), i.e. a change opposite in direction to the one produced by RLE. Additionally, they reported that in the case of chloroplasts in 10^{-5} M 3(3,4-dichlorophenyl)-1,1-dimethylurea (DCMU), the ratio F 735/F 693 also rose (to 2.9). No such change in ratio was observed in the present work, upon addition of CMU to chloroplasts. To determine whether the F 735/F 693 enhancement phenomenon were specific to DCMU, the experiment was repeated with several concentrations of DCMU, up to 10^{-4} M. It was not possible to reproduce the changes reported by Anderson and Thorne, unless the stock solutions of DCMU were prepared so as to yield final suspending media 3–5% ethanolic in water.

It may be recalled that in RLE-treated chloroplasts the "slow" changes in fluorescence induction exhibit a time dependence similar to that of conversion of the three-banded steady-state emission spectrum (77°K) into a (primarily) one-banded spectrum. From this, it seems likely that both these phenomena are brought about by action of the same component of the extract. Among the effects of the slower acting component on fluorescence induction is a marked diminution of the variable portion. Briantais (28), using system II and system I chloroplast fragments, isolated by differential centrifugation of detergent-treated chloroplasts, showed that the variable portion of fluorescence induction is greatly enriched and greatly reduced, respectively, in these fragments, compared to whole chloroplasts.

Since variable fluorescence is closely related to system II, it is not surprising that its marked diminution by the slower acting component of RLE is paralleled by decreases in both Hill activity (9, 29, 30) and system II-associated light-induced changes in absorption (24). Again, it may be worthwhile to compare the action of CMU with that of the slower acting component of RLE. Although CMU also inhibits system II, it does not block electrons from reaching Q. In contrast, one notes that with increasing time of exposure to RLE, less Q^- is produced. Thus, it would seem that the slower acting component of RLE is either blocking electron flow between H_2O and the reaction center of system II or is preventing electrons from getting to Q via the reaction centers. In the latter category are the possibilities of reaction centers being destroyed or electrons leaving the reaction centers but being detoured away from Q. (From oxygen transient studies (25), decrease in number of system II reaction centers was given as one of the effects of RLE on chloroplasts.) On the other hand, the diminution of variable fluorescence may result, not from aberrant photochemistry, but from impairment of energy transfer. Were the transfer of excitation energy to the reaction center chlorophylls diminishing because of disorganization of photosynthetic units, there would obtain a condition tantamount to decrease in number of functional reaction centers; that prolonged incubation of chloroplasts in RLE is accompanied by structural modifications (such as swelling of thylakoids and disruption of normal lamellar arrangement) has already been noted (9, 29, 30).

The observation was also made that prolonged exposure to RLE sometimes results in a strong retardation of *IP*, the slow component of the variable rise. Perhaps this retardation is a secondary effect which comes about because the interaction between Q^- and A has become relatively more efficient. At the present time, no explanation can be offered for the loss of some of the initial RLE-induced increase in amplitude of the fast component of variable fluorescence.

From experiments on the size and nature of the active components of RLE, particularly those with the two different fractions obtained through Sephadex chromatography, the following (see also reference 31) may be derived. The fast changes in fluorescence induction are brought about by the fraction of lower molecular weight, while the slower changes are brought about by the fraction of higher molecular weight; the latter (protein-like fraction) is also responsible for the changes in steady-state emission.[3] The experiments with intact *Chlorella* cells in RLE lend support to

---

[3] It has been noted that several substances, including long chain unsaturated fatty acids serve as good model systems for the action of the protein fraction of RLE (29, 30); that the changes in steady-state emission brought about by these substances result from a deaggregation of chlorophyll was

these contentions: the changes in fluorescence induction which occurred were almost exclusively of the "fast" type, and additionally, very little modification in steady-state emission (77°K) occurred. Such results could well be interpreted in relation to penetration of a small molecule, and exclusion of a large molecule, by the intact plasma membrane of the *Chlorella* cells.

Both components of RLE are of interest because they are naturally occurring substances that bring about some effects similar to those of extensively used synthetic compounds. Also, as seen in experiments on steady-state emission, the slower acting component is able to bring about large changes in the pathway of absorbed light energy by acting (directly or indirectly) on the chloroplast lamellae. It is possible that this molecule, or one closely related to it, represents the in vivo monitor for controlling both the partition of energy and electron transfer between system II and system I of photosynthesis.

The author wishes to express her appreciation to Professor Jean Lavorel for so graciously making available his laboratory facilities and for reading the manuscript.
She wishes also to thank Dr. H. de Klerk for his aid in chromatography experiments.

The experimental work reported here was done by the author at the Laboratoire de Photosynthèse, Centre Nationale de la Recherche Scientifique Gif-sur-Yvette, France, during sabbatical leave.

This work was supported, in part, by a CUNY Faculty Research Grant.

REFERENCES

1. Litvin, F. F., A. A. Krasnovskii, and G. T. Rikhireva. 1960. *Dokl. Akad. Nauk S.S.S.R.* **135**:1528.
2. Brody, M., and H. Linshitz. 1961. *Science (Washington)*. **133**:705, Fig. 1.
3. Bergeron, J. 1963. *Nat. Acad. Sci. Nat. Res. Counc. Publ.* **1145**:527.
4. Brody, S. S., and M. Brody. 1963. *Nat. Acad. Sci. Nat. Res. Counc. Publ.* **1145**:455.
5. Govindjee. 1963. *Nat. Acad. Sci. Nat. Res. Counc. Publ.* **1145**:318.
6. Kok, B. 1963. *Nat. Acad. Sci. Nat. Res. Counc. Publ.* **1145**:45.
7. Goedheer, J. C. 1964. *Biochim. Biophys. Acta.* **126**:304.
8. Broyde, S. B., and S. S. Brody. 1965. *Biochem. Biophys. Res. Commun.* **19**:444.
9. Brody, M., B. Nathanson, and W. S. Cohen. 1969. *Biochim. Biophys. Acta.* **172**:340.
10. McCarty, R., and A. Jagendorf. 1965. *Plant Physiol.* **40**:725.
11. Jagendorf, A., and M. Avron. 1958. *J. Biol. Chem.* **231**:277.
12. Lavorel, J., and C. Lemasson. 1961. *Biochim. Biophys. Acta.* **49**:574.
13. Arnon, D. I. 1949. *Plant Physiol.* **24**:1.
14. Lavorel, J. 1965. *Photochem. Photobiol.* **4**:819.
15. Lavorel, J. 1964. *Biochim. Biophys. Acta.* **88**:20.
16. Joliot, P., and J. Lavorel. 1964. *Bull. Soc. Chim. Biol.* **46**:1607.
17. Duysens, L., and H. Sweers. 1963. Studies on microalgae and photosynthetic bacteria. Japan Society of Plant Physiology, University of Tokyo, Tokyo. 353.
18. Joliot, P., A. Joliot, and B. Kok. 1968. *Biochim. Biophys. Acta.* **153**:635.

---

proposed by Nathanson and Brody (32). Experiments now in progress indicate that certain polyphenolics (e.g. rutin) induce changes in fluorescence induction similar to those brought about by the fast acting component of RLE.

19. Delosme, R. 1967. *Biochim. Biophys. Acta.* **143**:108.
20. Joliot, P. 1965. *Biochim. Biophys. Acta.* **102**:116.
21. Forbush, B., and B. Kok. 1968. *Biochim. Biophys. Acta.* **162**:243.
22. Munday, J., and Govindjee. 1969. *Biophys. J.* **9**:1.
23. Murata, N., M. Nishimura, and T. Takamiya. 1966. *Biochim. Biophys. Acta.* **120**:23.
24. Brody, M., S. Brody, and G. Döring. 1970. *Z. Naturforsch.* **25b**:862.
25. Brody, S. 1970. *Z. Naturforsch.* **25b**:855.
26. Lavorel, J. 1959. *Plant Physiol.* **34**:204.
27. Anderson, J., and S. Thorne. 1968. *Biochim. Biophys. Acta.* **153**:819.
28. Briantais, J. 1966. *Photochem. Photobiol.* **5**:135.
29. Brody, M. 1969. *Biophys. Soc. Annu. Meet. Abstr.* **9**:A-121.
30. Cohen, W. S., B. Nathanson, J. E. White, and M. Brody. 1969. *Arch. Biochem. Biophys.* **135**: 21.
31. Brody, M. 1969. Abstracts of the 11th International Botany Congress. Seattle. 22.
32. Nathanson, B., and M. Brody. 1970. *Photochem. Photobiol.* In press.

AN ACIDIC POLYSACCHARIDE FROM THE CELL WALL OF *CHLORELLA PYRENOIDOSA*

R. C. WHITE AND G. A. BARBER

SUMMARY

An acidic polysaccharide was solubilized from the cell wall of *Chlorella pyrenoidosa* with 4 M KOH in the presence of $NaBH_4$. It was purified by ion-exchange chromatography and gel filtration. Ultracentrifugal analysis of the preparation indicated that it was homogeneous with an approximate molecular weight of 88000. Upon complete acid hydrolysis the following monosaccharides were released: rhamnose, arabinose, xylose, mannose, galactose and glucuronic acid in molar ratios of 11.5:2.7:2.2:1:2.9:1.9. By use of specific enzymes it was demonstrated that rhamnose is present as the L-enantiomorph and galactose as a mixture of the D- and L-enantiomorphs in a ratio of 73:27. An aldobiuronic acid, glucuronosyl rhamnose, was isolated after partial acid hydrolysis of the purified polysaccharide.

INTRODUCTION

It has been observed that enzymes extracted from the unicellular green alga *Chlorella pyrenoidosa* catalyze the synthesis of UDP-L-rhamnose[1], probably GDP-L-galactose[2], and UDP-D-glucuronic acid (R. C. White, unpublished observation) as well as several other common sugar nucleotides[3]. The presence of such enzymes suggests that there may be polysaccharides in that organism containing one or more of the monosaccharides synthesized from those nucleotide derivatives. The possible involvement of D-glucuronic acid and L-galactose seemed particularly interesting since neither has been reported to occur in *Chlorella* polysaccharides.

When cell wall polysaccharides were extracted from *Chlorella* cells and fractionated by their solubilities in alkali, an acidic fraction could be isolated which yielded L-rhamnose, D-glucuronic acid, D- and L-galactose, arabinose, xylose and mannose upon acid hydrolysis. This paper will be concerned with the purification and partial characterization of that compound. Its relationship to the structure of the total cell wall will also be considered.

MATERIALS AND METHODS

Paper chromatography and electrophoresis

Partition chromatography was performed on Schleicher and Schuell No. 589

Blue Ribbon paper in the following solvent systems: Solvent 1, *n*-propanol–ethyl acetate–water (7:1:2, by vol.); Solvent 2, ethyl acetate–formic acid–acetic acid–water (18:3:1:4, by vol.).

Unless otherwise indicated, sugars were located on the chromatogram with the alkaline A_gNO_3 reagent[4].

High-voltage electrophoresis was carried out on a flat plate apparatus[5] at about 25 V/cm or in the Savant Model LT-20A tank unit at 80 V/cm on Schleicher and Schuell No. 589 Orange Ribbon paper. The buffer system used was either 0.1 M ammonium formate buffer (pH 3.7), or 0.05 M sodium borate (pH 9).

Gel filtration and ion-exchange chromatography

Bio-Gel P-300 (Calbiochem) was swollen for 24 h in 0.05 *M* Tris–HCl–1 M NaCl buffer (pH 8.3). Fine particles were removed by decanting the suspension several times. A 2.5 cm × 50 cm column (275 ml bed volume) was packed by gravity flow, and a small amount of Sephadex G-25 was layered on the top. It was equilibrated with 500 ml of the Tris–NaCl buffer before use.

Sepharose 6B (Calbiochem) was diluted to a thin slurry with 0.05 M Tris–HCl–1 M NaCl (pH 8.3) and packed by gravity flow into a 1.5 cm × 43 cm column (50 ml bed volume). The column was washed with 1 l of the Tris–NaCl buffer.

DEAE-Sephadex A-25-120 (Pharmacia) was swollen for 2 h in 0.05 M Tris–HCl (pH 8.3) and then packed by gravity flow into a 1.5 cm × 35 cm column (40 ml bed volume). The column was equilibrated with 500 ml of 0.05 M Tris–HCl (pH 8.3).

DEAE-cellulose was swollen in 1 M NaOH, washed with water, 0.5 M HCl, and the fines were then removed by decantation of the suspension. A 5 cm × 25 cm column was packed by gravity flow and equilibrated with 1 l of 0.005 M sodium phosphate buffer before use.

Culture of algae

Chlorella pyrenoidosa Chick strain 395 was obtained from the Culture Collection of Algae, Department of Botany, University of Indiana, Bloomington. It was maintained on agar slants of a salts medium[6] at 5°C. Subcultures were produced on the salt agar slants by incubation at room temperature for 2 weeks with continuous illumination 15 cm from two 25-W Gro-Lux lamps (Westinghouse). The cells from one such slant were used to inoculate aseptically 2 l of sterile salts medium[6] in a "Microferm" fermentor (New Brunswick Scientific Co.). The culture was thermostated at 28°C, gassed through a sparger at 3 l/min with 1.5% CO_2 in air and agitated with a rotary impellor at 180 rev./min. Light was supplied by a submerged water-cooled quartz lamp (General Electric T-3; λ_{max} 120 nm). The jacketed lamp housing was sterilized by immersion overnight in 800 ppm benzalkonium chloride ("Roccal"; Winthrop Laboratories) and washed with sterile water before insertion into the fermentor. The culture was incubated under continuous illumination for 4 days, after which 10 l of fresh medium were added to the fermentor. Continuous illumination was then continued until the cell density reached about $3 \cdot 10^7$ cells/ml (about 3 days). The cells were harvested in a Sharples centrifuge, washed once with water, and lyophilized. The yield was about 0.09 g dry wt per l.

Extraction of cell wall polysaccharide

A suspension of 2.0 g of lyophilized cells in 150 ml 0.025 M sodium–phosphate

Chlorella cells (2 g dry wt)

Suspended in 150 ml of 0.025 M sodium–potassium phosphate (pH 7.0). Passed through a French cell at 15 000 lb/inch^2. Centrifuged at 160 x *g* for 20 min

Supernatant solution | Residue

Supernatant solution: Centrifuged at 20 000 x *g*, 20 min

Supernatant solution | Residue

Residue: Treated with 500 ml acetone for 6 h at room temperature
Centrifuged at 20 000 x *g* for 20 min

Supernatant solution | Residue

Residue: Treated with 500 ml methanol–chloroform (1:1, v/v) at room temperature.
Centrifuged at 20 000 x *g* for 20 min

Supernatant solution | Residue

Residue: Suspended in 4 vol. ethanol

Supernatant solution | Insoluble material (lyophilized, yield 200 mg)

Insoluble material: Suspended in 20 ml of 0.1 M sodium–potassium phosphate (pH 7.0), 1 ml of human salivary amylase added, and the mixture was incubated overnight at 30°C. Added 4 vol. of 95% ethanol

Supernatant solution | Precipitate

Precipitate: Suspended in 20 ml of 0.025 M Tris–HCl (pH 7.3) containing 15 mg of Pronase. Incubated overnight at room temperature. Added 4 vol. of 95% ethanol

Supernatant solution | Precipitate (lyophilized, yield 100 mg, 5%)

Precipitate: Treated with 20 ml of 4 M KOH–0.03 M $NaBH_4$ at room temperature for 2 h while stirring with a stream of N_2 gas. Repeated extraction once. Centrifuged at 20 000 x *g* for 20 min

Alkali-soluble polysaccharide (46 mg, 2.3%) | Residue (14 mg, 0.7%)

Alkali-soluble polysaccharide: Brought pH to 8.3 with cold 50% acetic acid. Dialyzed solution against three 1-l volumes of 0.05 M Tris–HCl (pH 8.3) for 24 h at 4°C

Final solution of polysaccharides in 0.05 M Tris–HCl (pH 8.3)

Fig. 1. Details of the procedure used to extract polysaccharide from the cell wall of *Chlorella pyrenoidosa*.

buffer (pH 6.8) was passed two times through a French pressure cell at 15000–20000 lb/inch^2. Whole cells were removed from the homogenate by centrifugation at 160 $\times$ g for 20 min. The disrupted cell wall material was obtained by centrifugation at 20000 $\times$ g for 20 min. Pigments and other non-polar components were removed from the material by solvent extraction. Starch was removed by treatment with α-amylase and subsequent removal of the degraded glucan by dialysis. Protein was similarly removed by treatment with pronase. The material remaining was fractionated by extraction with 4 M KOH containing 0.03 M $NaBH_4$ under a stream of N_2. The alkali-soluble material was brought to pH 8.3 with acetic acid and dialyzed against several 1-l changes of buffer. This material was stored in solution at 4°C. The yield of alkali-soluble polysaccharide from 2 g dry wt of cells was 46 mg (2.3%) (determined by lyophilization of an aliquot). Details of the extraction and fractionation procedures are given in Fig. 1.

Gas–liquid chromatography

Monosaccharides obtained from polysaccharide hydrolysates were reduced to the alditols and acetylated as described by Albersheim *et al.*[7]. Small quantities of these were then injected into a 0.3 cm $\times$ 120 cm glass column containing 0.4% silicone, 0.2% ethylene glycol succinate, and 0.2% neopentyl glycol adipate on "adiatoport S" (80/100) in a Hewlett-Packard Model 402 gas chromatograph. At injection the flash heater was 200°C and the flame detector was 250°C. The oven was preheated to 110°C and held at this temperature for 6 min after injection, then a rise of 1°C/min was continued to 180°C. The monosaccharides were identified and quantitated by comparison of their positions and the areas under their peaks to known quantities of authentic compounds.

Analytical enzymes

L-Rhamnose isomerase was prepared by the method of Wilson and Ajl[8] and used as described previously[9]. D-Galactose dehydrogenase[10] was purchased from Sigma. L-Fucose dehydrogenase, which also reacts with L-galactose, was prepared from fresh pork liver through the DEAE-cellulose stage[11]. The dehydrogenase activities were monitored by following the increase in optical absorbancy at 340 nm of NADH in the reaction mixtures.

Reagents

L-Galactose and D-rhamnose were kindly supplied by Dr. Nelson Richtmyer. L-Rhamnulose was prepared by the method of Wilson and Ajl[8]. All other sugars were the purest available from commercial sources.

RESULTS

Purification of polysaccharide

The alkali-soluble polysaccharides from 2 g of lyophilized cells were applied to the DEAE-Sephadex column, and the column was washed with six bed volumes (240 ml) of 0.05M Tris–HCl buffer (pH 8.3). Polysaccharides were eluted with a 400 ml linear NaCl gradient (0 to 1 M) at a flow rate of 25 ml/h. 5-ml fractions were collected, and an aliquot of each was assayed for carbohydrate by the phenol–H_2SO_4 method[12].

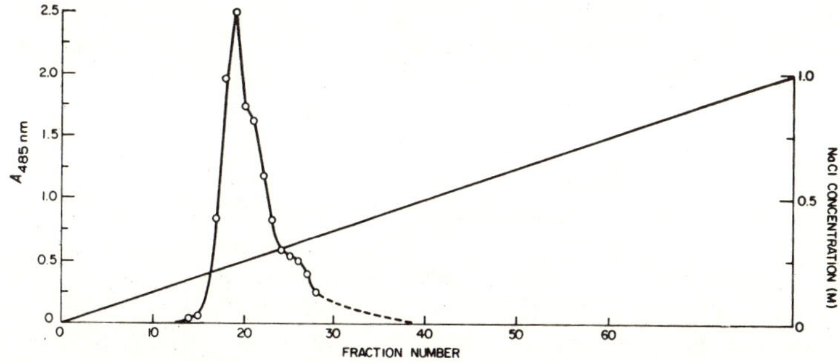

Fig. 2. Elution pattern from DEAE-Sephadex of the alkali-soluble polysaccharides of *Chlorella pyrenoidosa* cell wall.

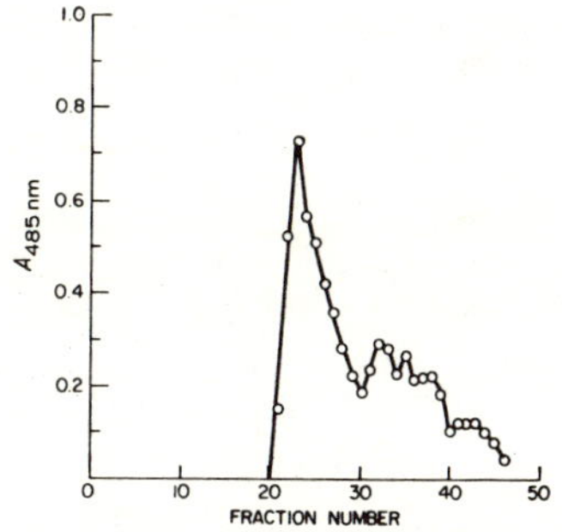

Fig. 3. Fractionation of the partially purified polysaccharide on a column of Bio-Gel P-300. The procedure is described in the text.

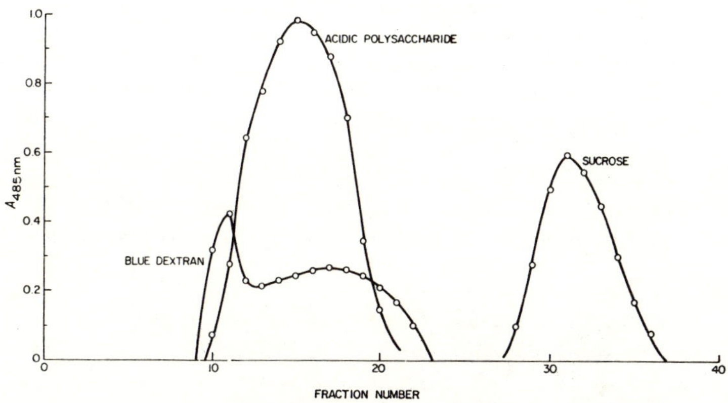

Fig. 4. Chromatography of the purified acidic polysaccharide on a column of Sepharose 6B. Exclusion limit for a polysaccharide on this column is $1 \cdot 10^6$. Blue dextran (2 mg/ml; av. mol. wt. $2 \cdot 10^6$) and sucrose (2 mg/ml) were added as reference compounds. Details of the procedure are given in the text.

The elution profile is shown in Fig. 2. Fractions 15–23 were pooled and dialyzed against 1 l of 0.05 M Tris–HCl buffer (pH 8.3) for 24 h at 4°C. The dialyzed solution was reduced to 10 ml in a rotary evaporator at 35° and was stored in the refrigerator.

A 2-ml sample of the material obtained from the DEAE-Sephadex column was made 2 M with NaCl and applied under the layer of buffer, through a long hypodermic needle, to the top of a column of Bio-Gel P-300. The column was eluted with 300 ml of 0.05 M Tris–HCl–1 M NaCl buffer (pH 8.3). Fractions of 3 ml were collected, and an aliquot of each was assayed for carbohydrate by the phenol–H_2SO_4 method[12]. The fractionation effected is shown in Fig. 3. Material in the first peak was collected and dialyzed against 1 l of 0.05 M Tris–HCl buffer (pH 8.3) for 24 h at 4°C. The volume of the dialysate was reduced to 3 ml at 35°C, and it was stored in the refrigerator.

The dialyzed Bio-Gel eluate was made 2 M in NaCl and applied to a column of Sepharose 6B. Polysaccharide was eluted from the column with 200 ml of Tris–HCl–1 M NaCl buffer (pH 8.3) at a flow rate of 36 ml/h. The elution pattern is shown in Fig. 4. The acidic polysaccharide appeared as a single symmetrical peak. Fractions containing the compound were pooled and dialyzed against several 1-l volumes of distilled water for 36 h at 4°C. This solution was stored at —15°C and is hereafter referred to as the purified polysaccharide.

Ultracentrifugal sedimentation properties of the polysaccharide

A solution containing 0.75 mg/ml of the purified polysaccharide was made 0.2 M in NaCl and centrifuged at 40000 rev./min in a Beckman Model E analytical ultracentrifuge. The polysaccharide sedimented as a single symmetrical peak (Fig. 5) and its sedimentation coefficient was calculated to be 2.07 S (ref. 13). The molecular weight of the compound was estimated by the sedimentation equilibrium procedure of Yphantis[14]. This was carried out as follows: A solution containing 0.5 mg/ml of the purified polysaccharide was made 0.1 M in NaCl and 0.03 M in NaH_2PO_4 and centrifuged at 14000 rev./min. Equilibrium was attained after 24 h. The run was continued for a total of 48 h. Data from a photographic plate of the Rayleigh interference pattern, plotted as ln of the fringe displacements (f) against the radial distance squared (r^2) in cm (Fig. 6) gives a straight line with a slope of 1.2974 (by Chi-squares method). This slope and a value for the partial specific volume (0.65), obtained experimentally by Olaitan and Northcote[15] for a similar polysaccharide, were used to calculate an approximate molecular weight of 88000.

Monosaccharide substituents of the polysaccharide

Lyophilized purified polysaccharide (2 mg) was dissolved in 2 ml of 2 M trifluoroacetic acid and hydrolyzed for 1 h at 100°C. The hydrolysate was transferred

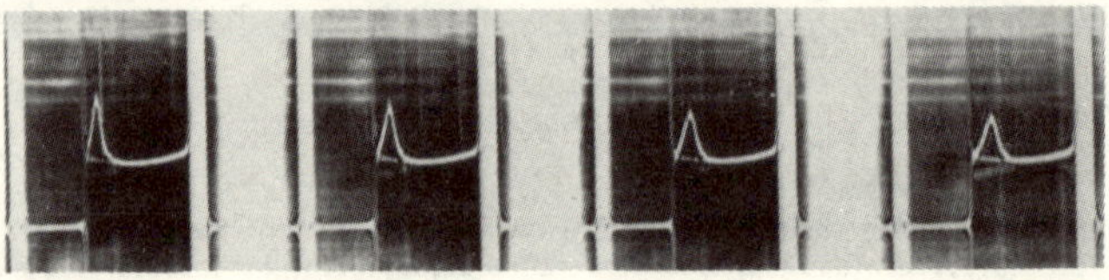

Fig. 5. Schlieren patterns of the purified polysaccharide sedimented centrifugally as described in the text.

to a 50-ml round bottom flask and evaporated to dryness on a rotary evaporator at 35 °C. The concentration of acid does not rise during the evaporation since the boiling point of trifluoroacetic acid (69 °C) is lower than that of water. Isoinositol (0.2 mg) was added as an internal standard, and the mixture was reduced and acetylated as described by Albersheim *et al.*[7].

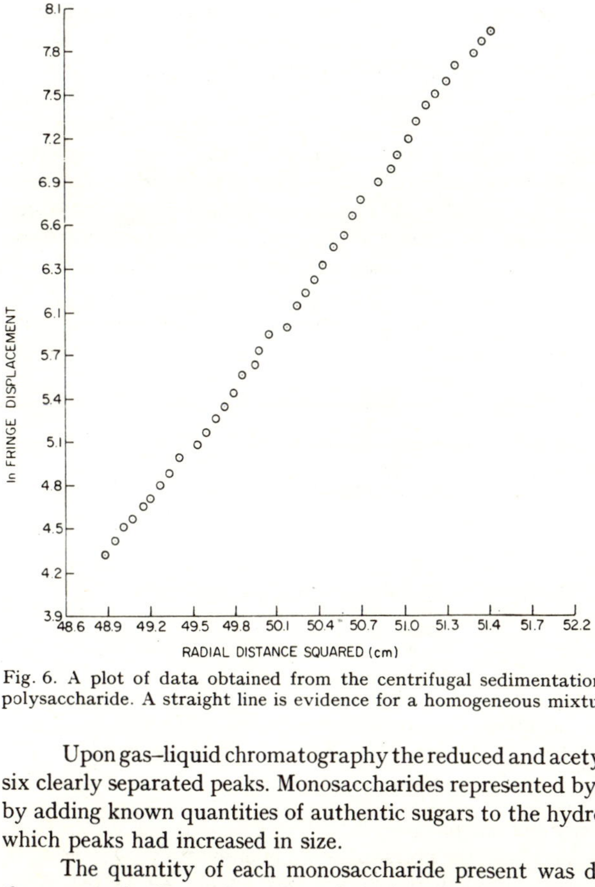

Fig. 6. A plot of data obtained from the centrifugal sedimentation equilibrium of the purified polysaccharide. A straight line is evidence for a homogeneous mixture of polymers[14].

Upon gas–liquid chromatography the reduced and acetylated hydrolysate yielded six clearly separated peaks. Monosaccharides represented by the peaks were identified by adding known quantities of authentic sugars to the hydrolysis mixture and noting which peaks had increased in size.

The quantity of each monosaccharide present was determined by comparing the area under its peak to that of the internal standard, isoinositol. An adjustment was made for the difference in yield of acetylated sugar alcohol from each monosaccharide[7].

Although a volatile derivative of glucuronic acid could be produced by reduction, methylation of the carboxylic group and acetylation, the yield of the derivative was not reproducible enough to give reliable quantitative data. Therefore, the proportion of glucuronic acid in the polysaccharide was estimated by the method to follow. A 3-mg sample of the lyophilized polysaccharide was hydrolyzed in 2 M trifluoroacetic acid at 121 °C for 2 h. Trifluoroacetic acid was removed *in vacuo*, and the residue was taken up in a small volume of water and electrophoresed at pH 3.7 at about 80 V/cm for 20 min. The bulk of the material was applied as a band on the paper, and a small portion was applied as a spot at a slight distance from the band. After electrophoresis the area of the paper that had been spotted was cut out and sprayed with the $AgNO_3$ reagent. Material in two areas of the electrophoretogram reacted with the $AgNO_3$ reagent. One area was at the origin (neutral sugars) and the other was indistinguishable in mobility from a glucuronic acid standard. The neutral sugar and uronic acid areas of the paper were cut out and eluted with water. The quantity of glucuronic acid present in the eluate was estimated by the method of Bitter and Muir[16]. A small amount of radioactive D-galactose was added as a marker, and the mixture of neutral sugars was chromatographed in Solvent 1 for 15 h. Galactose, which is well separated from the other monosaccharides under those conditions, was located by its radioactivity, eluted from the paper with water, and its quantity estimated by the method of Park and Johnson[17]. In each of the two methods an eluate of a blank portion of the paper was subjected to the same colorimetric procedure, and the results were used to correct for reactive materials in the paper, buffer and solvent system. The molar ratio of galactose to glucuronic acid was estimated to be 1:0.66. From this ratio and the data obtained by gas–liquid chromatography, the molar ratios and the percent weight of the monosaccharides in the polysaccharide were estimated. They are given in Table I.

TABLE I

ANALYSIS OF MONOSACCHARIDES IN THE PURIFIED POLYSACCHARIDE

Monosaccharide present	*Molar ratio*	*Percent weight of original polysaccharide**
Rhamnose	11.5	47.0
Arabinose	2.7	10.3
Xylose	2.2	8.2
Mannose	1	4.5
Galactose	2.9	12.8
Glucuronic acid	1.9	8.5

* These calculations do not include the increase in weight due to the addition of water during hydrolysis. The complete structure of the polysaccharide would have to be known before this could be accurately calculated.

The configuration of galactose and rhamnose

Galactose and rhamnose from the polysaccharide hydrolysate were isolated by chromatography in Solvent 1. The purity of each was established by gas–liquid chromatography.

L-Rhamnose isomerase[8] was used to determine the configuration of the isolated rhamnose. The reaction mixture contained 20 μl of enzyme (approx. 0.4 mg protein), 0.1 μmole neutral glutathione, 0.1 μmole $MnCl_2$, 0.2 μmole of the isolated rhamnose and 0.25 μmole Tris–HCl buffer (pH 7.5) in a total volume of 30 μl. Two other reaction mixtures were prepared as controls, one with 0.2 μmole authentic D-rhamnose and the other with 0.2 μmole authentic L-rhamnose. The mixtures were incubated for 30 min at 37°C, then applied to a paper moistened with 0.05 M sodium borate buffer (pH 9) and electrophoresed in that buffer. Under those conditions the authentic

L-rhamnose and rhamnose from the polysaccharide hydrolysate were converted almost entirely into a compound indistinguishable from authentic L-rhamnulose. D-Rhamnose was unaffected by the enzyme.

The proportion of galactose in the polysaccharide present as the D-enantiomorph was determined by the use of D-galactose dehydrogenase[10]. NADH produced in the following reaction mixture was found to increase linearly with increasing concentrations of D-galactose; 3 μl Sigma D-galactose dehydrogenase, 0.5 μmole NAD$^+$, 4 μmoles Tris–HCl buffer (pH 8.6) and 0.005–0.015 μmole D-galactose in a total volume of 0.3 ml. The mixtures were incubated for 30 min at 37°C. The absorption of NADH at 340 nm was used as a measure of the oxidation of the sugar. In a reaction mixture which contained both authentic D-galactose and galactose isolated from the polysaccharide, the reduction of NAD$^+$ was not inhibited and the results were additive. By this method the galactose sample from the polysaccharide was estimated to contain 126 μg/ml of D-galactose. The quantity of total reducing sugar in the sample was estimated as 173 μg/ml by the method of Park and Johnson[17]. Thus 73 % of the galactose in the polysaccharide is present as the D-enantiomorph. Since the sample contained only galactose according to the evidence of gas–liquid chromatography, it appeared likely that the remaining 27 % of reducing sugar represented the L-enantiomorph.

In order to obtain positive evidence for the presence of L-galactose, the sample was incubated with L-fucose dehydrogenase partially purified from pork liver. That enzyme has been shown also to oxidize L-galactose[11]. The reaction mixture contained: 1 ml of L-fucose dehydrogenase (approx. 0.1 mg protein eluted from DEAE–cellulose), 2.5 μmoles NAD$^+$, 50 μmoles Tris–HCl buffer (pH 8.3) and 1 μmole of galactose, in a total volume of 1.1 ml. Reaction mixtures were prepared using authentic L-galactose, authentic D-galactose and galactose isolated from the polysaccharide hydrolysate.

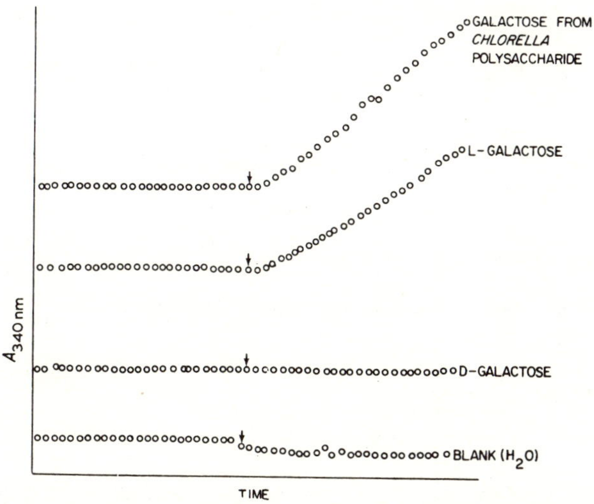

Fig. 7. Tracing of the recorder chart obtained upon incubating various sugars with L-fucose dehydrogenase. At the points indicated by the arrows substrate was added to the reaction mixtures. The distance between each point on the curves is 40 s.

The reactions were carried out at room temperature in 1-ml quartz cuvettes. The optical absorption of NADH at 340 nm was followed continuously with a Guilford Model 240 recording spectrophotometer. The results shown in Fig. 7 indicate that L-galactose is present in the algal polysaccharide. Unfortunately this assay could not be made quantitative because of colored impurities in the enzyme preparation and the small amount of monosaccharide available.

Isolation of an aldobiuronic acid from the polysaccharide

Lyophilized polysaccharide from the DEAE-Sephadex step (Fig. 2) (50 mg) was dissolved in 12 ml of 0.75 M trifluoroacetic acid, sealed in an ampoule and heated for 4 h at 100°C. A small amount of precipitate that formed during heating was removed by centrifugation. The supernatant solution was evaporated to dryness *in vacuo*, and the residue was taken up in water and evaporated to dryness again. This procedure was repeated three times. The residue was then dissolved in 5 ml of water and applied to a 1.5 cm × 37 cm column of the anion exchange resin AG-1-X8 formate. The column was washed with 1 l of water, and acidic components were then eluted with a 400-ml linear gradient of formic acid (0–2 M). Fractions of 3 ml were collected at a flow rate of 40 ml/h and assayed for carbohydrate by the phenol–H_2SO_4 method[12]. Three clearly defined peaks, labeled A, B, and C, were observed (Fig. 8). The material in each peak was collected and subjected to paper electrophoresis and

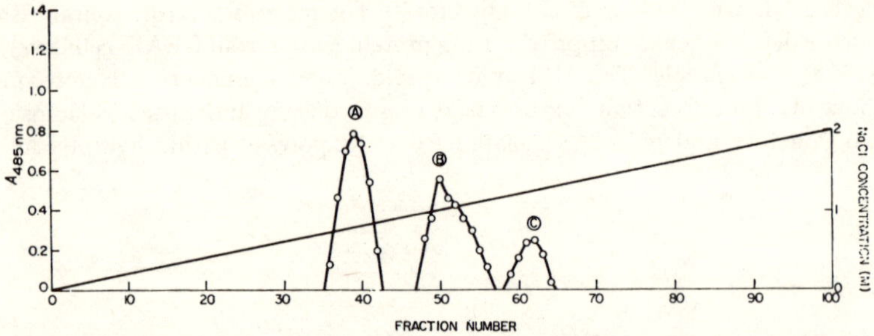

Fig. 8. Separation of the acidic components from a hydrolysate of the purified polysaccharide. The material was applied to a column of anion exchange resin and eluted as described in the text.

TABLE II

CHROMATOGRAPHY AND ELECTROPHORESIS OF THE URONIC ACID-CONTAINING PRODUCTS OF PARTIAL ACID HYDROLYSIS

Migration of the compounds are expressed relative to the migration of glucuronic acid (R_{ga}).

Fraction	R_{ga}	
	Electrophoresis at pH 3.7	*Chromatography in Solvent 2*
A	0.05	0.2
B_1	0.80	0.25
B_2	0.87	0.80
C	1	1

chromatography. The results are shown in Table II. The rates of migration of the materials in Fractions B1 and B2 suggested that they were aldobiuronic acids. Upon reduction of their free aldehyde groups with $NaBH_4$, methylation of the carboxyl groups with methanolic HCl (5 % HCl in methanol, 16 h at room temperature), hydrolysis (2 M trifluoroacetic acid, 2 h, 100°C), reaction of the monosaccharides with $NaBH_4$ and acetylation of the resultant sugar alcohols, the fractions were analyzed by gas–liquid chromatography. Compound B1 yielded peaks representing approximately equimolar quantities of rhamnose, glucuronic acid and mannose with a trace of xylose. Compound B2 yielded only glucuronic acid and rhamnose in almost equimolar quantities. We interpret these results to indicate that Compound B2 is an aldobiuronic acid of the structure glucuronosyl-L-rhamnose and that B1 probably consists of a mixture of two glucuronosyl disaccharides, glucuronosyl rhamnose (with a linkage different from B2) and glucuronosyl mannose. The xylose is thought to be a contaminant derived from the filter paper. In each case glucuronic acid presumably represents the non-reducing member of the disaccharide pair since the glucuronosyl hemiacetal bonds in a polysaccharide are usually the most resistant to acid hydrolysis[18].

When Fraction C was similarly treated, the results of gas–liquid chromatography indicated that it was free glucuronic acid. Fraction A, probably a large oligosaccharide, was present in quantities too small to permit identification of its components.

DISCUSSION

In this work the purification of a heteropolysaccharide from the cell wall of *Chlorella pyrenoidosa* is described. It appears to be a relatively homogeneous preparation since on ultracentrifugation it gave a single symmetrical peak and on equilibrium sedimentation a straight-line relationship of ln fringe displacement against radial distance squared was obtained (Fig. 6)[14].

We feel that the most notable features of this polysaccharide are the presence of glucuronic acid and both D- and L-galactose. Neither L-galactose nor glucuronic acid has heretofore been isolated from *Chlorella*, and, in fact, it has been especially noted that no uronic acid or L-galactose could be isolated from one strain of this species[15,19]. It also seems unusual to have found such a high proportion of L-rhamnose (almost 50 %) in the polysaccharide. It is possible that environmental conditions as well as constitutive factors in the strain may cause considerable variation in cell wall composition. In regard to L-galactose, it is of interest that an enzyme system that brings about its synthesis *via* the GDP–D-mannose derivative has recently been extracted from the strain of *Chlorella pyrenoidosa* used in this investigation[2].

The approximate molecular weight determined for the compound by ultracentrifugal analysis (88000) is considerably lower than would be expected from the rate of its movement through the various molecular sieves. This may be attributable to the shape of the molecule. If it were highly branched, it should be more difficult for it than for a linear compound, such as a polypeptide, to pass through the pores of a filtration gel. Hence it would be eluted from the column more quickly than the linear compound.

NORTHCOTE *et al.*[19] have concluded that the *Chlorella* wall is composed of two phases, one an organized microfibrillar structure and the other a matrix in which it is embedded. The matrix could be removed and microfibrillar material isolated by treatment of wall preparations with dilute alkali[19]. Our results indicate that approximately 75 % of the alkali-soluble polysaccharide (the matrix) in this strain is composed of acidic polysaccharides and that a large proportion of that material is made up of the purified polysaccharide (approx. 60 %). It was also observed that both the non-acidic and acidic polysaccharides of the alkali-soluble fraction contain very similar proportions of neutral sugars. Thus, it is possible that the acidic polysaccharide represents a repeating unit of the cell wall matrix that was specifically removed from a continuous network during the isolation procedure.

The cell wall material that was insoluble in alkali yielded only glucose and glucosamine on acid hydrolysis. This suggests that it is unrelated to the soluble polysaccharides and is perhaps derived from a mixture of cellulose and a chitin-like substance. These polysaccharides probably make up the microfibrillar structure of this organism and account for the unusual durability of its cell wall.

ACKNOWLEDGMENTS

This research was supported in part by National Institutes of Health Grant No. 13267. R.C.W. is the recipient of a National Science Foundation Traineeship. Dr Kirk Aune and Mrs Linda Jones rendered much valuable advice and assistance in the use of the analytical ultracentrifuge.

REFERENCES

1 G. A. Barber and M. T. Y. Chang, *Arch. Biochem. Biophys.*, 118 (1967) 659.
2 G. A. Barber, *Arch. Biochem. Biophys.*, 147 (1971) 619.
3 H. Kauss and O. Kandler, *Z. Naturforsch.*, 176 (1962) 858.
4 L. Hough and J. K. N. Jones, *Methods Carbohydr. Chem.*, 1 (1962) 21.
5 A. M. Crestfield and F. W. Allen, *Anal. Chem.*, 27 (1955) 769.
6 A. Kuhl and H. Lorenzen, *Methods Cell. Physiol.*, 1 (1964) 159.
7 P. Albersheim, D. J. Nevins, P. D. English and A. Karr, *Carbohydr. Res.*, 5 (1967) 340.
8 D. M. Wilson and S. Ajl, *J. Bacteriol.*, 73 (1957) 410.
9 G. A. Barber, *Biochim. Biophys. Acta*, 165 (1968) 68.
10 K. Wallenfels and G. Kurz, *Methods Enzymol.*, 9 (1966) 112.
11 H. Schachter, J. Sarney, J. McGuire and S. Roseman, *J. Biol. Chem.*, 244 (1969) 4785.
12 J. E. Hodge and B. T. Hofreiter, *Methods Carbohydr. Chem.*, 1 (1962) 388.
13 H. K. Schachman, *Methods Enzymol.*, 4 (1957) 55.
14 D. A. Yphantis, *Biochemistry*, 3 (1964) 297.
15 S. A. Olaitan and D. H. Northcote, *Biochem. J.*, 82 (1962) 509.
16 T. Bitter and H. Muir, *Anal. Biochem.*, 4 (1962) 330.
17 J. T. Park and M. J. Johnson, *J. Biol. Chem.*, 181 (1949) 149.
18 K. K. De and T. E. Timell, *Carbohydr. Res.*, 4 (1967) 177.
19 D. H. Northcote, K. J. Goulding and R. W. Horne, *Biochem. J.*, 70 (1958) 391.

RELATIONSHIPS BETWEEN THE PHOTON DISTRIBUTION BETWEEN THE TWO PHOTOSYSTEMS, THE CONCENTRATION OF SYSTEM II REACTION CENTERS AND THE INTERSYSTEM EQUILIBRIUM CONSTANT IN *CHLORELLA PYRENOIDOSA*

M.-J. DELRIEU AND Y. DE KOUCHKOVSKY

SUMMARY

From Emerson enhancement measurements of O_2 evolution in *Chlorella pyrenoidosa*, it was possible to establish a relationship between the concentration of photosystem II open reaction centers (E) and the distribution of photons between photosystems I and II $[(1-\alpha)/\alpha]$ during steady state. The superposition of lights of two different wavelengths (1 and 2) gives concentrations of E and α intermediate between those obtained with light 1 and 2 separately. This relationship extends a previous one based on quantum yield measurements. It has been expressed here by a curve corresponding to a fixed value of the intersystem apparent equilibrium constant (K). Up to 700 nm, K remains equal to 6. Above this wavelength, although the margin of error is rather great, K apparently increases to 12 or more.

The possibility of "spill-over" of light absorbed by System II to System I was studied. There is no probability that this spill-over, if any, exceeds 25 % in Chlorella.

The apparent equilibrium constant is decreased by 3(3,4-dichlorophenyl)-1,1-dimethylurea. This is not in favor of the hypothesis of fully independent electron-transfer chains in photosynthesis; it is therefore likely that some communication between those chains exists.

INTRODUCTION

JOLIOT *et al.*[1] have found that it was possible to determine an equilibrium constant (from 3 to 10) between the concentration of photochemical centers of System I (P_{700}) and System II (E or Q) in isolated chloroplasts. Likewise with Chlorella, quantum yield values (from 600 to 680 nm) could be interpreted by assuming an equilibrium constant equal to 6 (ref. 2) which suggests the occurrence of a back flow of electrons.

As a matter of fact, however, the equilibrium constant K so defined probably does not describe a true thermodynamic situation. Redox potential determination

Abbreviations: CMU, 3(*p*-chlorophenyl)-1,1-dimethylurea; DCMU, 3(3,4-dichlorophenyl)-1,1-dimethylurea.

of fluorescence[3] and P_{700} (ref. 4) suggest a much higher equilibrium constant ($K \simeq 2000$).

On the other hand, MALKIN[5] has shown that the observed K can be accounted for by a model of independent electron-transfer chains. This allows the simultaneous coexistence of reaction centers in oxidized and in reduced forms; they do not react with one another because they belong to different chains. We consider the intersystem equilibrium constant to be a characteristic of the relationship between the two photochemical centers that does not necessarily explain their coupling mechanism.

The quantum yield experiment, published in a preceding paper[2], allowed us to determine the equilibrium constant with precision in a small range of wavelengths, in which the two systems absorb in nearly equal quantities (between 600–680 nm). We propose to determine K at longer wavelengths from the Emerson enhancement values and in particular to study whether or not it remains constant.

MATERIAL AND METHODS

The biological material is the unicellular green alga *Chlorella pyrenoidosa* Chick, Emerson strain, grown under controlled conditions on synthetic mineral medium (macroelements of TAMIYA *et al.*[6], microelements A_4 and B_7 of ARNON[7] *plus* $6 \cdot 10^{-5}$ M EDTA).

O_2 exchange was measured using a bare concentration-type electrode (the cathode and anode are directly in contact with the medium). This electrode was similar to that of JOLIOT[8] but adapted here for quantum yield measurements by placing an annular photoresistor directly under the cuvette containing the electrodes.

The cuvette (Fig. 1) is cylindrical (25 mm in diameter, height 4 mm). It includes in its inner surface a thin layer of platinum, the edge of which forms the cathode, and the reference electrode is a thick layer of silver, plated with chloride. These electrodes are separated by Plexiglas. The medium is stirred with a Plexiglas disc, the diameter of which is 0.2 mm less than that of the cuvette. Its shaft passes through a Teflon bearing in the bottom of the cuvette; it is driven by a synchronous motor at 3000 turns/min. The equilibration of the disc must be excellent, so that the liquid in the cuvette runs in a laminar manner along the electrodes (vibration creates whirls and noise in the recording). The cathode is polarized at -0.7 V. The current variations are amplified and registered with a "Graphispot" (Sefram) recording galvanometer (period: 0.45 sec).

The cuvette can be illuminated with two 500-mm Bausch-and-Lomb monochromators.

The rates of O_2 evolution are measured by the differences in the recorded slopes, just before and just after the end of illumination. The rapid response of the galvanometer eliminates in these measurements, interference caused by "respiration changes" of the algae.

The standard experimental conditions are the following: Chlorella at about 200 μg chlorophyll per ml (always the same concentration for a given set of experiments), suspension equilibrated with air, at 20°, illumination over 1000 ergs$\cdot$cm$^{-2}\cdot$sec^{-1}, depending on the experiment and the wavelength (but always well below the saturation); average half band width, approx. 6.5 nm.

(A) The model used

(a) Definitions. e_{I}, e_{II}, fractions of the traps of photosystem I and photosystem II in the open form; (E = photosystem II traps); k_{I}, k_{II}, quantum yield of photoreactions I and II, respectively; i, total light quanta absorbed; α, fraction of the total absorbed light which is absorbed by photosystem II; i_{I}, i_{II}, light absorbed by photosystem I and photosystem II: $i_{I} = (1-\alpha)i$, $i_{II} = \alpha i$, thus

$$\frac{i_{I}}{i_{II}} = \frac{1-\alpha}{\alpha} \tag{1}$$

v_{I}, v_{II}, rates of photoreaction I and of O_2 evolution (photoreaction II), respectively.

(b) Relationship between v_{II} and e_{II}. For photoreaction II, JOLIOT AND JOLIOT[9] found a relationship between the rate of O_2 evolution, v_{II}, and the concentration of photosystem II open centers, e_{II}. This relationship is expressed by:

$$v_{II} = k_{II} i_{II} \frac{e_{II}}{1-p+e_{II}p} \tag{2}$$

(p is the probability factor of energy transfer between System II traps $\simeq 0.5$). We set:

$$u = \frac{e_{II}}{1-p+e_{II}p} \tag{3}$$

so we obtain:

$$v_{II} = k_{II} i_{II} u \tag{4}$$

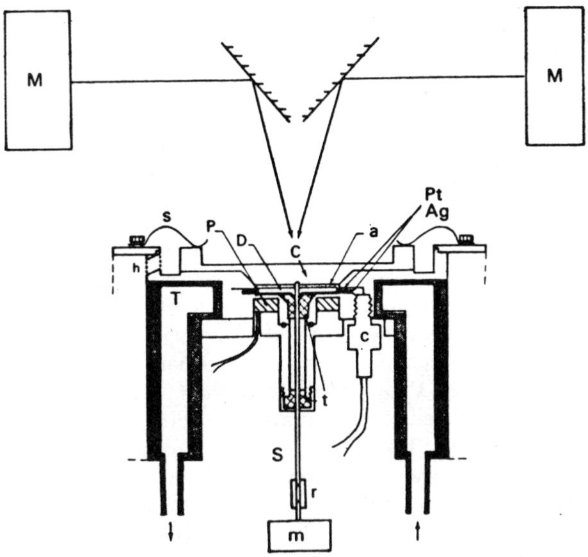

Fig. 1. Diagram of the apparatus for measuring O_2 exchanges and light absorption by algae. Pt, Ag, platinum and silver (chloride plated) electrodes; a, algal suspension; C, cover; D, stirring disc; S, shaft; P, photoresistor; s, spring; h, hole for evacuating the used suspension; T, thermoregulated bath; t, Teflon bearing; c, coaxial connector; m, synchronous motor; r, rubber tubing (connecting stirring disc and motor shaft); M, monochromator.

(*c*) *Equilibrium between photoreactions I and II active centers*. Photoreaction I re-creates open centers of photosystem II (*i.e.* E), with a rate,

$$v_I = k_I i_I e_I \tag{5}$$

Eqn. 5 is based on the assumption that no energy migration exists between System I traps (see ref. 1).

During steady-state O_2 evolution, under light-limiting conditions, the variation of e_{II}, which is equal to its rate of formation by photoreaction I *minus* its rate of destruction by photoreaction II, is zero (neglecting possible dark redox reactions on E):

$$k_I i_I e_I - k_{II} i_{II} u = 0 \tag{6}$$

From Eqn. 1 we find therefore:

$$\frac{k_I}{k_{II}} \cdot \frac{1-\alpha}{\alpha} = \frac{u}{e_I} \tag{7}$$

We suppose that e_I and e_{II} are linked together by an equilibrium constant K:

$$\frac{(1-e_I)(1-e_{II})}{e_I e_{II}} = K^{-1} \tag{8}$$

so,

$$e_I = \frac{1-e_{II}}{1+(K^{-1}-1)e_{II}}$$

and with Eqn. 3:

$$e_I = \frac{1-u}{1+Au} \tag{9}$$

where $A = K^{-1}(1-p)-1$. Therefore Eqn. 7 becomes:

$$\frac{k_I}{k_{II}} \cdot \frac{1-\alpha}{\alpha} = u\,\frac{1+Au}{1-u} \tag{10}$$

(*d*) *Effect of spill-over*. The above calculation is developed assuming that there is no "spill-over"[10] from photosystem II to photosystem I. It is more complete if we try to introduce a simple model of direct excitation transfer from System II to System I. This model does not involve the physical nature of the medium in which the excitation diffuses (effect of three-dimensional space[11], heterogeneity, *etc.*) and implies that all other types of energy dissipation (such as heat) remain unchanged.

The excitation is created in System II by photons with the rate αi. According to Joliot and Joliot[9], each excited closed reaction center, the total concentration of which is $(1-e_{II})$, has a probability p to transfer its excitation to another open or closed center in System II. The probability for n transfers between $(n+1)$ centers is $p^n(1-e_{II})^{n+1}$. But, competitively, the energy may also be transferred from a closed photosystem II center to System I; the probability of this spill-over is defined as r (since p was found $\simeq 0.5$, ref. 9, $r \leqslant 0.5$). Then, the probability that System I will receive an excitation from System II after n transfers in System II is $rp^n(1-e_{II})^{n+1}$. Once the excitation is transferred to System I, the probability of its utilization is $k_I e_I$. Therefore, the part of the System I rate that would be due to spill-over is equal to:

$$v_I(II \rightarrow I) = \sum_n \alpha i p^n (1 - e_{II})^{n+1} r k_I e_I$$
$$= k_I \alpha i e_I r \frac{1 - e_{II}}{1 - p + e_{II} p} \qquad (11)$$

At the steady state, v_I, the total photoreaction I rate due to the absorption of photons by System I and to the spill-over, is equal to v_{II}, the photoreaction II rate:

$$k_{II} \alpha i \frac{e_{II}}{1 - p + e_{II} p} = k_I (1 - \alpha) i e_I + k_I \alpha i r e_I \frac{1 - e_{II}}{1 - p + e_{II} p} \qquad (12)$$

Our model does not imply that the transfer of excitation to photosystem I has an effect of decreasing the rate of photoreaction II by the effective diminishing of α as was supposed by MALKIN[12]. We consider instead that the effect of spill-over in adjusting the rates of the two photoreactions is through the adjustment of the concentrations of e_1 and e_2 (*i.e.* u). Therefore, with Eqn. 3, Eqn. 12 becomes:

$$\frac{k_I}{k_{II}} \frac{1 - \alpha}{\alpha} = u \frac{1 + Au}{1 - u} - \frac{k_I}{k_{II}} \cdot \frac{r(1 - u)}{1 - p} = D(u) \qquad (13)$$

The higher the number of inactive centers of System II $(1 - u)$, the greater is the effect of the spill-over (by comparing with Eqn. 10, r and p being constant). So at 700 nm, $(1 - u)$ is small and the effect of spill-over is negligible; on the contrary, at 650 nm it will be maximal (if it exists).

We want to obtain the experimental curve $D(u) = [k_I/k_{II}]\,[(1 - \alpha)/\alpha]$ and compare it to the preceding theoretical relationship (Eqn. 13).

(B) Relationship between the curve D(u) and the Emerson enhancement effect.

(I) Expressions of Emerson enhancement

If algae are illuminated with two light intensities, i_1 and i_2, of wavelengths λ_1, λ_2, we obtain: v_1, steady-state rate of O_2 evolution with $i_1(\lambda_1)$; v_2, steady-state rate of O_2 evolution with $i_2(\lambda_2)$; v_{12}, steady-state rate of O_2 evolution with $i_1 + i_2(\lambda_1, \lambda_2)$.

We can define the enhancement by:

$$E_1 = \frac{v_{12} - v_2}{v_1} = \frac{v_{12} - v_1 - v_2}{v_1} + 1 \qquad (14)$$

$$E_2 = \frac{v_{12} - v_1}{v_2} = \frac{v_{12} - v_1 - v_2}{v_2} + 1 \qquad (15)$$

$(E_1 - 1)v_1 = (E_2 - 1)v_2$, therefore

$$E_2 - 1 = \frac{v_1}{v_2}(E_1 - 1) \qquad (16)$$

(see Fig. 2). We may also define enhancement in another manner by:

$$E_{12} = \frac{v_{12}}{v_1 + v_2} \qquad (17)$$

$$E_{12} - 1 = (E_2 - 1)\frac{v_2}{v_1 + v_2} = (E_1 - 1)\frac{v_1}{v_1 + v_2} \qquad (18)$$

If we set

$$x = \frac{v_2}{v_1 + v_2} \qquad (19)$$

plotting $E_{12} - 1$ as a function of x, the slope of the curve at $x = 0$ is $E_2^{max} - 1$, and for $x = 1$ it is $1 - E_1^{max}$ (see Fig. 3).

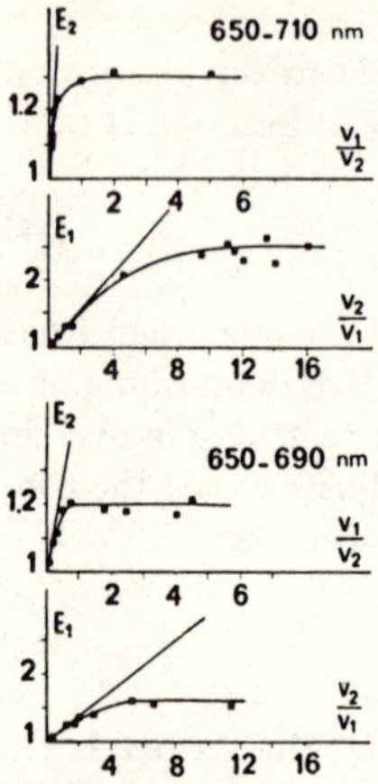

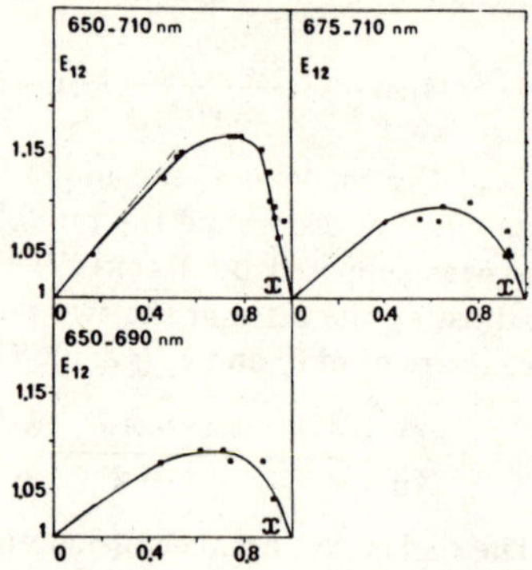

Fig. 2. Enhancement calculated as E_1 *versus* v_2/v_1 ratio and E_2 *versus* v_1/v_2 ratio for the given pairs of wavelengths. Standard experimental conditions (see text).

Fig. 3. Enhancement as E_{12} *versus* $x = v_2/(v_1 + v_2)$ for the given pairs of wavelengths. Same experiment as in Fig. 2.

(II) Construction of the curve D(u) with the Emerson enhancement

The O_2 rate when the algae are illuminated with two wavelengths, λ_1 and λ_2, separately is:

$$\text{For } \lambda_1: v_1 = k_{II}\alpha_1 i_1 u_1$$

$$\text{For } \lambda_2: v_2 = k_{II}\alpha_2 i_2 u_2 \qquad (20)$$

(Eqn. 12 shows that these expressions are true even if spill-over exists.) If we use the two lights simultaneously ($i_1\lambda_1$, $i_2\lambda_2$) the new rate v_{12} is proportional to the new concentration of u. Since the total of photons absorbed is $\alpha_1 i_1 + \alpha_2 i_2$, $v_{12} = k_{II}(\alpha_1 i_1 + \alpha_2 i_2)u$. From Eqn. 20,

$$k_{II}\alpha_1 i_1 = \frac{v_1}{u_1} = \rho_1 \text{ and } k_{II}\alpha_2 i_2 = \frac{v_2}{u_2} = \rho_2 \qquad (21)$$

Therefore:

$$v_{12} = (\rho_1 + \rho_2)u \qquad (22)$$

Since $i = i_1 + i_2$, the new value of α, the fraction of photons received by pigment System II, is:

$$\alpha = \frac{\alpha_1 i_1 + \alpha_2 i_2}{i_1 + i_2} \qquad (23)$$

(the general relationship $v = k_{II}\alpha i u$ is always true).

The expression $[k_I/k_{II}]\cdot[(1 - \alpha)/\alpha] = D(u)$, which is the curve to be studied, becomes, in the case where both wavelengths are used simultaneously:

$$\frac{k_{I}}{k_{II}}\cdot\frac{1-\alpha}{\alpha} = \frac{D_1\rho_1 + D_2\rho_2}{\rho_1+\rho_2} = \frac{D_1\dfrac{1-x}{u_1} + D_2\dfrac{x}{u_2}}{\dfrac{1-x}{u_1} + \dfrac{x}{u_2}} = D(u) \tag{24}$$

(Knowing Eqns. 23, 21, 19 and that

$$D_1 = \frac{1-\alpha_1}{\alpha_1}\cdot\frac{k_{I}}{k_{II}} \text{ at } \lambda_1 \text{ and } D_2 = \frac{1-\alpha_2}{\alpha_2}\cdot\frac{k_{I}}{k_{II}} \text{ at } \lambda_2.)$$

The different geometrical and mathematical demonstrations are based only on an experimental relationship, Eqn. 2. They do not contain any physical hypothesis of the process which gives the function $D(u)$.

(*a*) *Relationship between* E_{12} *and the shape of the curve* $D(u)$. In Eqn. 24, $D(u)$ is the average of D_1 and D_2 weighted, respectively, with ρ_1 and ρ_2 (see Eqn. 21) and therefore Eqn. 24 is equivalent to:

$$\frac{D(u)-D_1}{D(u)-D_2} = -\frac{\rho_2}{\rho_1} = \frac{u_c-u_1}{u_c-u_2} \tag{25}$$

where u_c refers to the abscissa of the chord joining D_1 to D_2 (see Fig. 4). From Eqn. 22:

$$v_{12} - v_1 - v_2 = (E_{12} - 1)\,(v_1 + v_2) = (u - u_1)\rho_1 + (u - u_2)\rho_2 \tag{26}$$

This expression, which represents the increase above the sum of rates $v_1 + v_2$, is the average of $(u - u_1)$ and $(u - u_2)$ weighted by the same respective factors ρ_1 and ρ_2.

If the curve $D(u)$ is a straight line (the chord) joining D_1 to D_2, Eqn. 26 is equal to 0. So, the Emerson effect is a measure of the curvature of the curve between D_1 and D_2.

If we set:

$$\delta u = u - u_c \tag{27}$$

the difference in the abscissa at the ordinate $D(u)$ between the curve and the chord joining D_1 and D_2, we find from a simple geometrical consideration of Fig. 4:

$$(E_{12} - 1)\,(v_1 + v_2) = \delta u(\rho_1 + \rho_2) = \delta u\left(\frac{v_1}{u_1} + \frac{v_2}{u_2}\right)$$
$$\text{and } E_{12} - 1 = \delta u\left(\frac{x}{u_2} + \frac{1-x}{u_1}\right) \tag{28}$$

Knowing x, u_1 u_2 and E_{12} by experiment, one may therefore obtain δu, with which the curve between u_1 and u_2 may be constructed (if D_1 and D_2 are known).

(*b*) *Relationship between* $E_1 - 1$ *and the curve* $D(u)$. From Eqns. 18 and 28 we have:

$$E_1 - 1 = (E_{12} - 1)\,\frac{v_1 + v_2}{v_1} = \delta u\,(\rho_1 + \rho_2)\,\frac{1}{v_1}$$

in the limit, $v_1 \to 0$ (this is taken generally in the experiments to find E_1: $v_1 \ll v_2$ and $\rho_1 \ll \rho_2$ therefore $u \to u_2$).

Recalling u_c, the abscissa of the chord corresponding to the ordinate $D(u)$; St_2, the slope of the tangent at $D_2 = [(D(u) - D_2)/(u - u_2)](u \to u_2)$; Sc_{12}, the slope of the chord from $D(u)$ to $D_2 = [(D(u) - D_2)/(u_c - u_2)]$; we find:

$$(E_1 - 1)v_1 \to 0 = \frac{u_1 - u_2}{u_1}\left(\frac{Sc_{12}}{St_2} - 1\right) \quad (29)$$

(*c*) *Experimental construction of the curve D*(u). At a first approximation, u_1 is considered as being equal to 1 (λ_1 close to 700 nm). The values of u_2 were measured precisely in the range $0.73 < u < 0.84$ in a previous work[2]; experimentally these values were obtained by the ratio $v_s/v_{i\ preil}$ at a given wavelength (v_s, steady-state rate of O_2 evolution; $v_{i\ preil}$, initial rate after a preillumination near 710 nm, $v_s = k_{II}\alpha i u$, $v_{i\ preil} = k_{II}\alpha i$).

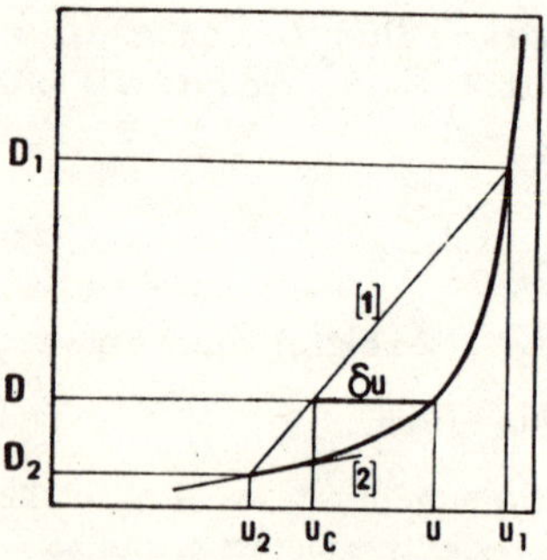

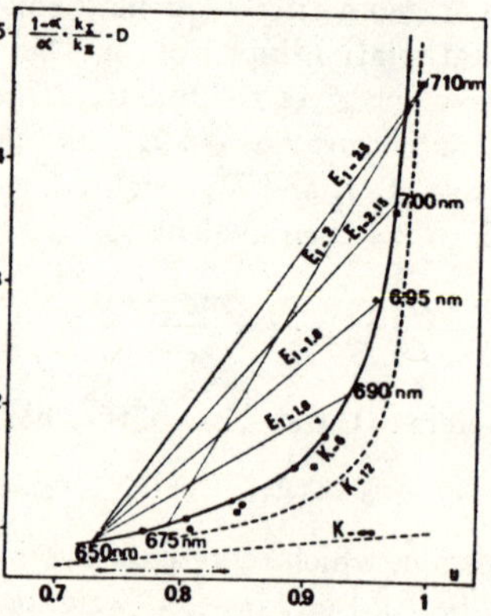

Fig. 4. Schematic representation of the curve $D(u)$. 1, chord joining D_1 and D_2; 2, tangent at D_2; δu, difference in the abscissa at the ordinate D between the curve $D(u)$ and the chord joining D_1 and D_2.

Fig. 5. Curve $[k_I/k_{II}]\,[(1-\alpha)/\alpha] = D(u)$. The slopes of the chords joining D_2 and D_1 are obtained from the E_1^{max} data (see text). The experimental points are obtained from the E_{12} data (●, 650–690 nm; ▲, 675–710 nm; ■, 650–710 nm). The curves are calculated for different values of the equilibrium constant K·(←·—·→, range of wavelengths previously studied from quantum yield measurements.[2])

From Eqn. 29, knowing u_1, u_2, St_2 and $E_1^{max} - 1$, we can find the value of Sc_{12}. This allows us to draw the chord joining D_1 to D_2. (St_2, the slope at the tangent at D_2, is measured graphically on the curve; the data pertaining to the curve $D(u)$ for $0.73 < u < 0.84$ have been previously obtained[2]. In Fig. 5, one can see that the intersection of two chords corresponding to two different values of u_2 (λ_2 being 675 nm instead of 650 nm for instance) permits the calculation of a more precise value of D_1.

Knowing the chord, with Eqn. 28, one may build the whole curve between D_2 and D_1; the values of E_{12} as a function of x give a series of values of δu that are drawn from the chord at the corresponding values of the ordinate given by Eqn. 24.

RESULTS AND DISCUSSION

(*A*) *Emerson enhancement measurements*

The Emerson enhancement values obtained with Chlorella are given in Figs. 2 and 3. We first tried to use the results from the literature[10,13], but they were not complete enough to be used with confidence in this work. These authors did not measure the ratio $v_s/v_{i\ \text{preil}} = u$ (for the concentration of E) for the wavelengths λ_1 and λ_2. It is therefore not possible to verify the similarity of our material with that of MYERS AND GRAHAM[10] for example. The curve $E_{12}(x)$ was drawn only for 700 nm by ELEY AND MYERS[13] and should be subject to caution in interpretation. For instance, the absorptions of lights 1 and 2 being different ($A_{650\ \text{nm}} \simeq 3 \times A_{700\ \text{nm}}$) the intensities of λ_1 and λ_2 do not vary in the same ratio through the thickness of the layer if a rate electrode is used. The same phenomenon can also occur with the concentration electrode, although it is minimized by the rapid stirring of the suspension.

Fig. 5 gives the curve obtained by the method described above from the experimental results of Emerson enhancement alone. Therefore it gives the distribution of photons $(1-\alpha)/\alpha$ between System I and II (assuming $k_I = k_{II}$) as a function of the steady-state concentration of the photosystem II traps. The experimental points fit on a theoretical curve obtained using an equilibrium constant K equal to 6 and a probability of spill-over equal to 0 (identical therefore to that obtained with the quantum yield values). However, the fit is not precise for $\lambda > 700$ nm. The equilibrium constant seems to vary here (it may be for instance 12 at 710 nm), but the margin of error for its determination is great (see Fig. 5).

(B) Study of the possibility of excitation transfer from System II to System I

We want to see if we can find agreement with $r > 0$ between the experimental curve and theoretical curves. For low values of $r(\leqslant 0.1)$ it is obvious that within the experimental errors, the agreement remains rather good. But if we try to find the value of K which fits the experimental curve for a higher value of r (r = 0.3, for example) it is impossible.

Fig. 6 shows this result. If $r = 0.3$, we may with Eqn. 13 substract from the experimental Curve 1 the quantity, $[k_I/k_{II}]\,[r(1-u)/(1-p)]$; we then obtain Curve 2, which would be the experimental curve if there was no spill-over. The curve thus

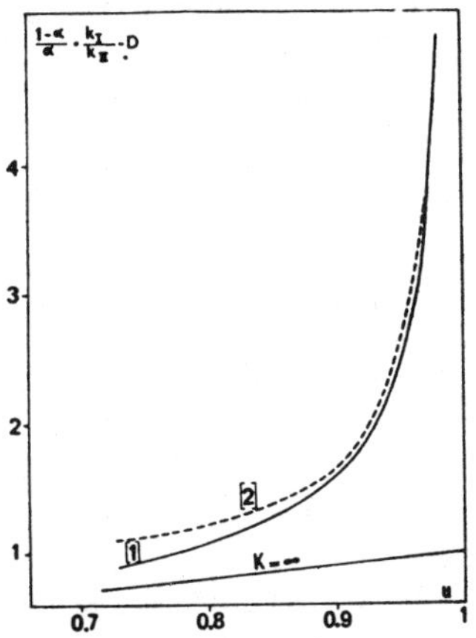

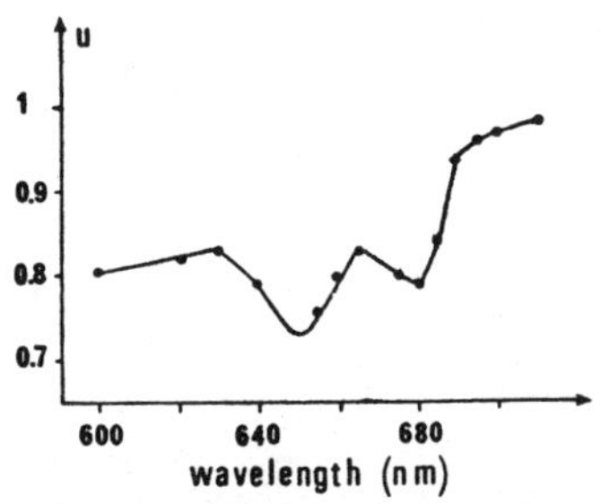

Fig. 6. Spill-over possibility. 1, experimental curve $D(u) = [k_I/k_{II}]\,[(1-\alpha)/\alpha]$; 2, theoretical curve without spill-over if it is supposed that for Curve 1 a spill-over of 30% exists.

Fig. 7. Variation of u (which measures the concentration of photosystem II traps) as a function of illumination wavelength $[u = e_{II}/(1-p+e_{II}p) = v_s/v_{i\ \text{preil}}]$.

obtained no longer corresponds to any positive equilibrium constant at $u = 0.73$ (650 nm) because the slope of Curve 2 at this point becomes less than the slope of the curve corresponding to $K = \infty$.

In addition, in order to find a theoretical curve fitting Curve 2 at points other than $u = 0.73$, it is necessary for the values of K to change continuously. For example, at 660 nm (corresponding to $u = 0.78$) K will be ∞, whereas at 700 nm $K = 6$. Thus a decrease of K is accompanying a decrease of α (*cf.* MALKIN[5]).

So our experimental results do not allow any possibility of spill-over greater than 25 % as it has been described here (we find a value of $K \leqslant \infty$ at $u = 0.73$ only when $r \leqslant 0.25$). However, even if, regarding the mechanism of energy transfer *per se*, the limit of 0.25 is not negligible with respect to the maximum possible value of $r(\simeq 0.5)$, the corresponding spill-over would have only a minor effect on the overall photosynthetic process.

(C) Considerations of the results

In Fig. 5, the values of the distribution of photons $(1-\alpha)/\alpha$ and of u are defined rather precisely at wavelengths between 685 and 710 nm, a range which it was not possible to study very well by direct quantum yield measurements[2]. Therefore, combining the results of these previous measurements (for $\lambda \leqslant 685$ nm) and of the present ones (for $\lambda \geqslant 685$ nm), one can rather accurately draw the curves u and $(1-\alpha)/\alpha$ *versus* λ (Figs. 7 and 8). From these curves, the quantum yield (Φ) values are readily obtained since $\Phi = k_{II}\alpha u$. In Fig. 9, the steady-state quantum yield spectrum obtained by EMERSON AND LEWIS[14] or SCHWARTZ[15] coincides with these values up to 700 nm. But the calculated value at 710 nm is distinctly higher than that measured experimentally by these authors. The reason for this discrepancv is not understood at the present time.

(D) Interpretation of our results with respect to the BONAVENTURA AND MYERS *effect*[16]

Continuous illumination by light absorbed mainly by photosystem I or II ("light 1" or "light 2") induces slow changes in the rate of O_2 evolution; in particular

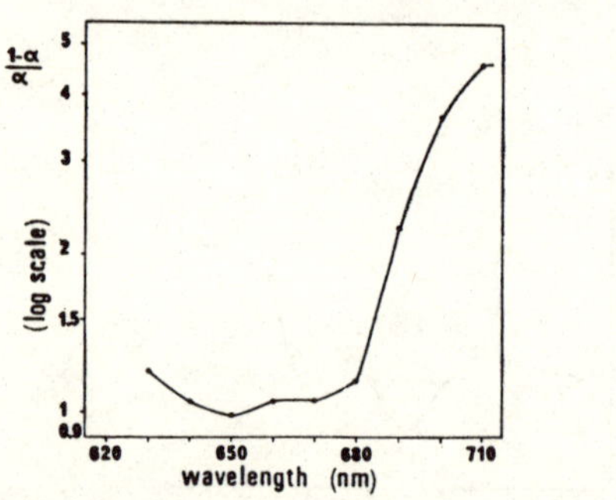

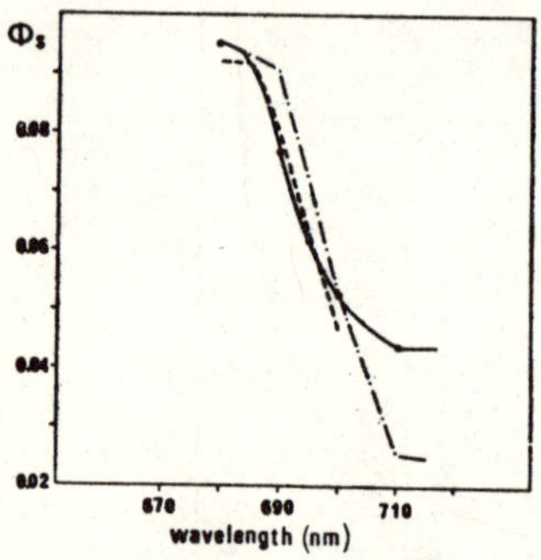

Fig. 8. Ratio between the fraction of light absorbed by System I $(1-\alpha)$ and the fraction of light absorbed by System II (α) *versus* wavelength.

Fig. 9. Wavelength profile of the quantum yield of steady-state photosynthesis from 680 nm. -----, O_2 evolution measured by EMERSON AND LEWIS[14]; –·–·–·, NADP+ reduction (relative quantum yields) measured by SCHWARTZ[15]; ———, calculated from u and α values obtained (see Figs. 7 and 8).

an increase of the quantum yield of about 15 % in 10 min when the algae, adapted to light 1 illumination (dark or prolonged illumination at 710 nm), are shifted to the light 2 state (by prolonged illumination at 650 nm)[16]. This phenomenon can be interpreted as a decrease of α in the transition from the light 1 state to the light 2 state; and inversely as an increase of α in the transition from the light 2 state to the light 1 state. It is thus necessary to define the conditions under which the experiments were made.

The measurement of u was made in state 1, the steady-state rate was measured after about 1 min of illumination. The value of u found here at 650 nm is therefore a true minimum; in state 2 it would increase like (E) or the quantum yield[16]. Emerson enhancement was also measured under the above conditions.

We can explain why, in the results of BONAVENTURA AND MYERS[16] during the transition to state 2 for example, the quantum yield (Φ_{O_2}) increases at 650 nm and decreases at 710 nm. By looking at Fig. 5 it can be seen that the decrease of α (increase of $(1-\alpha)/\alpha$) will result in a large increase of u (*i.e.* E) at 650 nm but a very small increase at 710 nm. Thus, if we consider the equation $\Phi_{O_2} = k_{II}\alpha u$, at 710 nm, α is the main factor which varies ($\alpha\downarrow, \Phi\downarrow$), whereas at 650 nm, u varies more than α, but in the opposite direction ($\alpha\downarrow, u\uparrow, \Phi\uparrow$).

(*E*) *Action of 3-(3,4-dichlorophenyl)-1,1-dimethylurea* (*DCMU*)

The inhibition site of DCMU is situated close to the photosystem II reaction centers[17]; DCMU prevents the oxidation of E by photoreaction I (refs. 17, 18) resulting in a decrease of E.

We have supposed that DCMU at the low concentration used does not affect the total absorption of photons (*i.e.* there is no change in the absorption spectrum[19]).

We assume that α and k_{II} (and more precisely the ratio $[k_I/k_{II}]$ $[(1-\alpha)/\alpha]$ stay identical to that of the uninhibited algae because it was shown[18] that the relationship $v_{O_2\uparrow} = k_{II}\alpha i u$ is the same whether a change of u is obtained by the action of 3-(p-chlorophenyl)-1,1-dimethylurea (CMU, an analog of DCMU) or by the action of different wavelengths of light.

We have obtained at different concentrations of DCMU ($1\cdot10^{-7}$–$4\cdot10^{-7}$ M) the

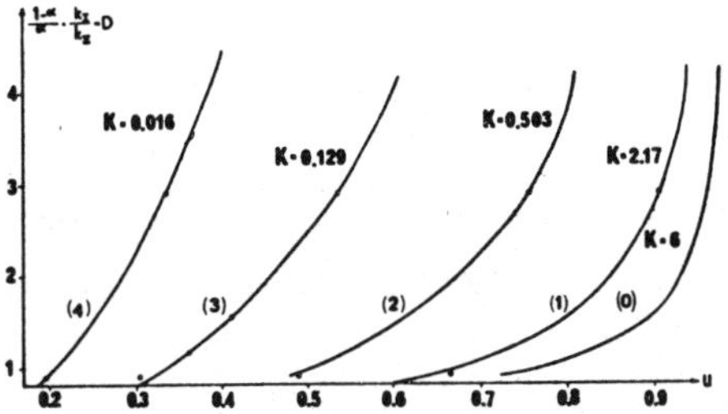

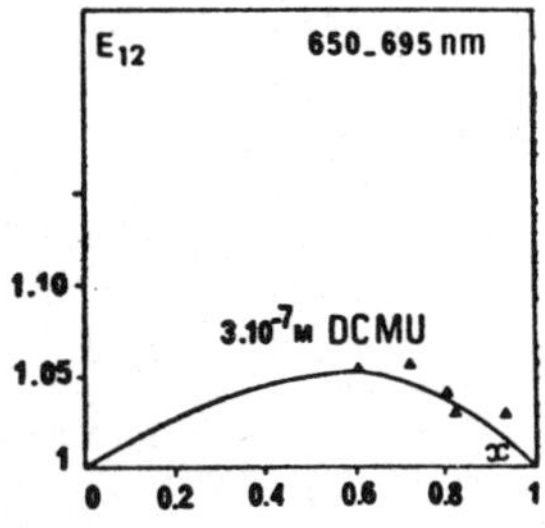

Fig. 10. Action of DCMU on the equilibrium constant K. 0, no DCMU (see Fig. 5); 1, $1\cdot10^{-7}$ M DCMU; 2, $2\cdot10^{-7}$ M DCMU; 3, $3\cdot10^{-7}$ M DCMU; 4, $4\cdot10^{-7}$ M DCMU. ●, direct measurements; ○, computed from Fig. 11. $u = v_s$ with inhibitor/$v_{1\ \mathrm{prell}}$ without inhibitor. Standard experimental conditions.

Fig. 11. Enhancement E_{12} *versus* $x = v_2/(v_1+v_2)$ with $3\cdot10^{-7}$ M DCMU. Standard experimental conditions.

new values of u when the algae are illuminated with 650- and 695-nm illumination [u = (v_s with inhibitor)/($v_{i\,prell}$ without inhibitor]. The decrease of u by DCMU is different according to the wavelength of illumination. This result is in agreement with those obtained by GINGRAS AND LEMASSON[17] that the action spectrum of CMU inhibition of O_2 evolution is that of photosystem II.

In Fig. 10, the ordinate $[k_I/k_{II}]$ $[(1-\alpha)/\alpha]$ is known for 650 and 695 nm. So, for each concentration of DCMU we can put points on theoretical curves corresponding to decreasing equilibrium constants. Furthermore, some measurements of Emerson enhancement effect have been performed in the presence of $3 \cdot 10^{-7}$ M DCMU. The enhancements E_{12} are low (Fig. 11). From these results, values of δu allow us to put points on the corresponding Curve 3 in Fig. 10.

On the one hand, if the photosynthetic chains are fully independent, the action of DCMU at a given moment on one chain will be total, whether or not the following equilibrium exists:

$$\text{active chain} + \text{DCMU} \rightleftharpoons \text{inhibited chain}$$

DCMU decreases the number of electron-transport chains, so the apparent equilibrium constant K should not change, since it is the result of statistical phenomena[5]. In this case k_I and k_{II} decrease in the same ratio: active chain/inhibited chain and there would be no change of k_I/k_{II}. (For the two photoreactions it is the same product, namely O_2, which is measured. Thus, blocking the chain at any point, the DCMU would affect in the same ratio the corresponding quantum yields k_1 and k_{II}.) The independent chain model is therefore in contradiction with the above experiment.

On the other hand, if there is a real equilibrium, as has been described in our model, there would also be no change of K by the addition of DCMU. The constant K would be independent of the total amount of one product of the chain as shown by Eqn. 8.

So, the relationship between the two systems is not rigorously an equilibrium. K is not really a constant; it is a function of the concentration of intermediate products (which are changed by DCMU). Therefore, we may propose, in conclusion, that a certain communication exists between the chains. It may be at the level of plastoquinone, as STIEHL AND WITT[20] have suggested to explain the mechanism of reduction or oxidation of plastoquinone.

REFERENCES

1 P. JOLIOT, A. JOLIOT AND B. KOK, *Biochim. Biophys. Acta*, 153 (1968) 635.
2 M.-J. DELRIEU, in H. METZNER, *Progress in Photosynthesis Research*, Vol. 2, H. Metzner, Tübingen, 1969, p. 1110.
3 B. KOK, S. MALKIN, O. OWENS AND B. FORBUSH, in *Energy Conversion by the Photosynthetic Apparatus, Brookhaven Symp. Biol.*, 19 (1966) 446.
4 B. KOK, *Biochim. Biophys. Acta*, 48 (1961) 527.
5 S. MALKIN, *Biophys. J.*, 9 (1969) 489.
6 H. TAMIYA, E. HASE, K. SHIBATA, A. MITUYA, T. IWAMURA, T. NIHEI AND T. SASA, in *Algal Culture from Laboratory to Pilot Plant, Carnegie Inst. Wash. Publ.*, 600 (1953) 204.
7 D. I. ARNON, *Am. J. Botany*, 25 (1938) 322.
8 P. JOLIOT, *Biochim. Biophys. Acta*, 102 (1965) 116.
9 A. JOLIOT AND P. JOLIOT, *Compt. Rend.*, 256 (1964) 4622.
10 J. MYERS AND J. R. GRAHAM, *Plant Physiol.*, 38 (1963) 105.
11 J. LAVOREL, *J. Chem. Phys.*, 47 (1969) 2235.
12 S. MALKIN, *Biophys. J.*, 7 (1967) 629.

13 J. H. Eley and J. Myers, *Plant Physiol.*, 42 (1967) 598.
14 R. Emerson and C. M. Lewis, *Am. J. Botany*, 30 (1943) 165.
15 M. Schwartz, *Biochim. Biophys. Acta*, 131 (1967) 559.
16 C. Bonaventura and J. Myers, *Biochim. Biophys. Acta*, 189 (1969) 366.
17 G. Gingras and C. Lemasson, *Biochim. Biophys. Acta*, 109 (1965) 67.
18 A. Joliot, *Physiol. Végétale*, 6 (1968) 235.
19 S. Izawa, in H. Metzner, *Progress in Photosynthesis Research*, Vol. 3, H. Metzner, Tübingen, 1969, p. 1742.
20 H. H. Stiehl and H. T. Witt, *Z. Naturforsch.*, 24b (1969) 1588.

Fatty Acid Metabolism of Green Algae

GAS CHROMATOGRAPHIC ANALYSIS OF THE HIGHER FATTY ACIDS OF THE ALGA *CHLORELLA VULGARIS* (*PYRENOIDOSA*)

M. MATUCHA, L. ŽILKA AND K. ŠVIHEL

SUMMARY

Fatty acids of the alga *Chlorella vulgaris* (*pyrenoidosa*) were analyzed by the gas–liquid chromatography of their methyl esters. By use of two (polar and non-polar) stationary phases, comparison with standards and bromination C_{14}–C_{20} fatty acids were detected and identified. In addition to the major acids (palmitic, oleic, linoleic and linolenic) the alga lipids contain *n*-14:0, 14:1, 14:2, 14:3, *n*-15:0, *ai*-15:0, 16:1, 16:2, 16:3, *n*-17:0, *ai*-17:0, *n*-18:0, *n*-19:0 and *n*-20:0 fatty acids at levels in excess of 0.1 %. The results obtained are of importance in the analytical control of the radiochemical purity of ^{14}C-labelled higher fatty acids prepared from the alga. The presence of odd (straight-chain and branched) acids is interesting from biochemical viewpoint.

INTRODUCTION

The lipids of *Chlorella vulgaris* (*pyrenoidosa*) contain four well-known major fatty acids (FA) of the C_{16} and C_{18} series (16:0, 18:1, 18:2, 18:3) and several minor components[1–5]. Not all of these minor components have been identified satisfactorily. A basic study in this field was made by MANGOLD AND SCHLENK[2], who found in *Chlorella pyrenoidosa* the following fatty acids: one 12:0, 14:0, 14:1, 14:2, 14:3, 15:0, 16:0, 16:1, 16:2, 16:3, two 17, one 18:0, 18:1, 18:2, 18:3, 19, three 20, two 22 and one 24. The attention of JAMES[6], HARRIS AND JAMES[7], NICHOLS[8,9], BARRON[10] AND MIYACHI[11] was drawn to the biosynthesis and metabolism of fatty acids. Other publications[12,13] contain few details.

In recent years a new technique has been used in the analysis of lipids: the combination of thin-layer (TLC) and gas–liquid chromatography (GLC). With this combined chromatographic technique whole lipid classes were analyzed (first separated by TLC and then chromatographed by GLC, *e.g.* ref. 14) or a detailed analysis of FA was performed. So KORN[15] analyzed the fatty acids of *Euglena gratilis*. He showed that some overlapping occurs during GLC and after elimination of this by use of TLC he was able to detect over fifty fatty acids.

EXPERIMENTAL

For the purpose of analysis of the algae and the control of radiochemical purity of ^{14}C-labelled fatty acids (prepared from radioactive algae[16]) we used a technique usually applied to the analysis of more complex mixtures of volatile compounds: GLC of methyl esters on two stationary phases. We tried to identify all *Chlorella* fatty acids with a relative abundance higher than 0.5 %. A polar polyester diethylene glycol succinate (DEGS) and a non-polar grease, Apiezon-L, were chosen as stationary phases.

Microorganisms

A pure culture of the alga *Chlorella vulgaris* No. 82 from the collection of autotrophic organism, Czechoslovak Academy of Sciences, Prague, precultivated as reported by BASLEROVÁ AND DVOŘÁKOVÁ[17], was grown under conditions similar to those for cultivation in an atmosphere of $^{14}CO_2$ (ref. 16).

Chemicals

Myristic, palmitic*, stearic, oleic, linoleic** and linolenic** acids were supplied by Lachema (ČSSR), Reanal (Hungary)* and Fluka (Switzerland)** and 14-methylhexadecanoic acid was obtained thanks to the kindness of Dr. J. HRADEC of the Oncological Institute, Prague.

Preparation of fatty acid methyl esters

After centrifugation of *Chlorella* cells and washing free of nutrient medium, lipids were extracted with hot 96 % ethanol and with an ethanol–diethyl ether mixture (3:1). The remaining part of the sugars was removed from the lipid extract by water extraction; the dried lipids were then transesterified with methanolic hydrochloric acid. After the extraction of methyl esters by petroleum ether they were purified by vacuum sublimation.

Gas–liquid chromatography

This was carried out on a column of length 2 m and I.D. 4 mm filled with 15 % DEGS on Chromaton N-AW-HMDS 60/80 (Lachema), in a Packard Chromatograph 7409 with an argon-ionisation detector (AID) and on a column of length 1.5 m and I.D. 3 mm filled with 15 % Apiezon-L on Chromaton N-AW-NMDS 60/80 (Lachema), in a Becker-Delft Chromatograph C-ASTGV with a flame-ionisation detector (FID).

For the labelled fatty acids the simultaneous determination of radioactivity, after combustion in Tricarb furnace Packard 325, was carried out by a 2-ml gas proportional counter. The counter unit consisted of a Berthold LB 80074 preamplifier with high voltage supply, an amplifier and a Berthold LB 242K dual-channel rate meter.

RESULTS AND DISCUSSION

We identified in the *Chlorella vulgaris* lipids *n*-fatty acids with even numbers of carbons saturated (14, 16, 18 and 20), mono-, di- and tri-unsaturated acids (14, 16 and 18) and *n*- and *anteiso*-fatty acids with odd numbers of carbons (15, 17, 19). As was expected, we found the overlapping[15] of peaks of *Chlorella* fatty acid methyl esters. This is obvious from the semi-logarithmic plots of retention times of the fatty acid methyl esters determined in *Chlorella versus* number of carbons in chain (Figs. 1 and 2). It is also evident from the GLC analysis of *Chlorella* fatty acids shown in Figs. 3 and 4.

The identification of individual compounds was carried out by comparison with standard substances or by bromination (the unsaturated acids then disappearing on GLC). The minor components, with the exception of 14-methylhexadecanoic acid which was isolated from lanolin, were identified using the regularities of their retention times. The position of the double bond in unsaturated acids was not verified and the position of the methyl group was verified only for 14-methylhexadecanoic acid (a component of carcinolipin) by its biochemical behaviour[18–20]. The other

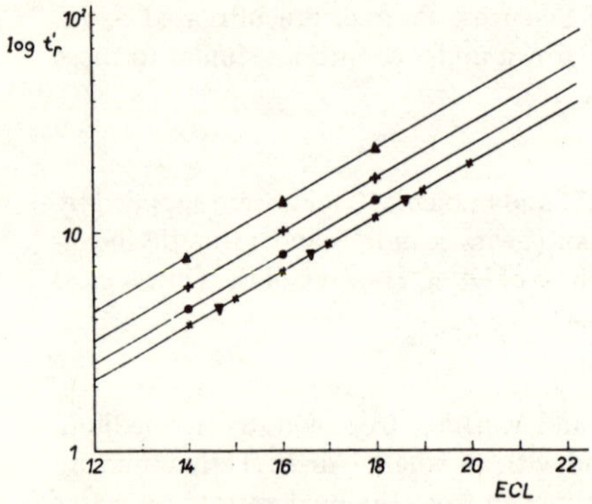

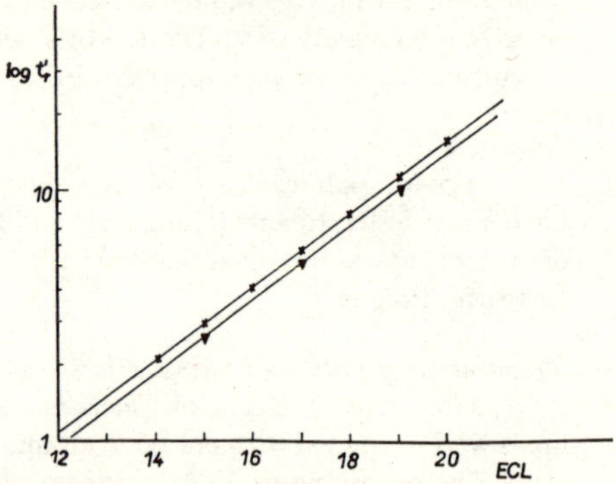

Fig. 1. The retention time of fatty acid methyl esters *versus* the equivalent chain length (polar phase). x, straight-chain saturated; ▼, *anteiso* saturated; ●, mono-unsaturated; +, di-unsaturated; ▲, tri-unsaturated.

Fig. 2. The retention time of fatty acid methyl esters *versus* the equivalent chain length (non-polar phase). ×, straight-chain saturated; ▼, *anteiso* saturated.

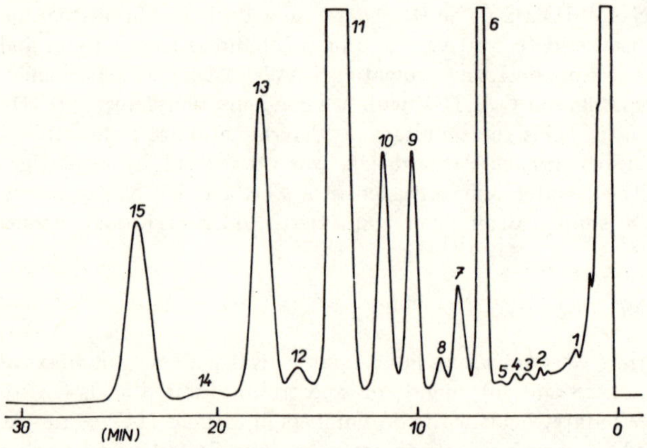

Fig. 3. Gas chromatogram of fatty acid methyl esters, prepared from triglyceride fraction of *Chlorella* lipids (TLC) on diethylene glycol succinate at 190°. Carrier gas: argon (30 ml/min). Argon ionisation detector. Column: length 2 m × 4 mm I.D. Esters: 1 = 12:0; 2 = 14:0; 3 = 14:1 and *ai*-15; 4 = 15:0; 5 = 14:2; 6 = 16:0; 7 = 16:1, *ai*-17 and 14:3; 8 = 17:0; 9 = 16:2; 10 = 18:0; 11 = 18:1, 16:3 and *ai*-19; 12 = 19:0; 13 = 18:2; 14 = 20:0; 15 = 18:3.

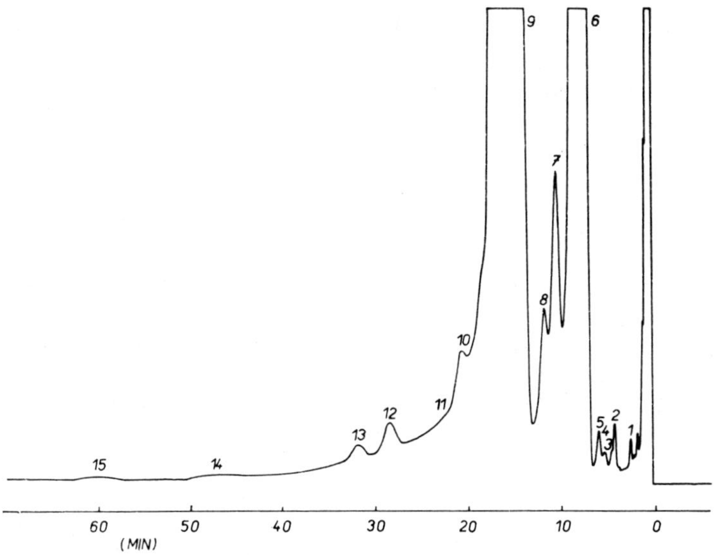

Fig. 4. Gas chromatogram of *Chlorella* fatty acid methyl esters on Apiezon-L at 230°. Carrier gas: nitrogen (30 ml/min). Flame ionization detector. Column: length 1.5 m × 3 mm I.D. Esters: 1 = 12:0; 2 = 14:1, 14:2 and 14:3; 3 = 14:0; 4 = *ai*-15; 5 = 15:0; 6 = 16:0, 16:1 and 16:2; 7 = *ai*-17; 8 = 17:0; 9 = 18:0, 18:1, 18:2 and 18:3; 10 = *ai*-19; 11 = 19:0; 12 = 20:1, 20:2; 13 = 20:0; 14 = 22:0; 15 = 23:0.

TABLE I

COMPOSITION OF *Chlorella* FATTY ACID METHYL ESTERS

Methyl ester of acid	*Weight* (%)
12:0	0.1
14:0	0.1
14:1	0.5
14:2	0.1
14:3	0.1
ai-15	0.1
15:0	0.5
16:0	15–18
16:1	0.5–1
16:2	3–5
16:3	10–12
ai-17	1–1.5
17:0	0.5
18:0	3–5
18:1	20–25
18:2	13–18
18:3	13–18
ai-19	0.2
19:0	0.1
20:0	0.1–0.2
20:1	0.2
20:2	0.2
higher acids than C_{20}	traces

anteiso-acids (Fig. 2) originate evidently from biosynthesis, as do the straight-chain odd- and even-numbered acids. The results obtained agree generally with the work of SCHLENK AND MANGOLD[2], who discussed the possibility of the presence of odd-numbered acids. We found these acids in the bacteria-free culture of *Chlorella* as well as in the ^{14}C-labelled fatty acids of *Chlorella vulgaris* and identified them more precisely. The quantitative composition of *Chlorella* fatty acids (Table I) differs insignificantly from published data[2]. The representative separation of *Chlorella* fatty acid methyl esters on DEGS is shown on a typical radio-gas chromatogram in Fig. 5. The composition of *Chlorella* fatty acids is important to the biosynthetic preparation of ^{14}C-labelled fatty acids and their isolation (combination GLC–TLC or two-phase GLC) and analysis. It is evident (Figs. 1 and 2) that for example in oleic acid separated on DEGS, 16:3 and *ai*-19:0 acids can also be present when TLC or separation on Apiezon-L or a similar phase was not carried out. This fact is naturally important for the choice of an isolation method.

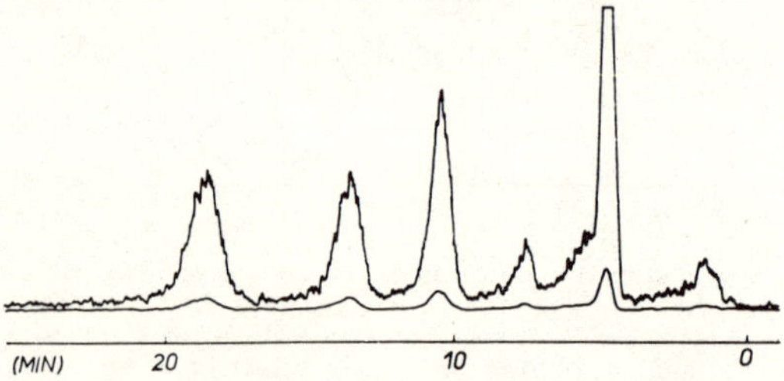

Fig. 5. Radio-gas chromatogram of *Chlorella* fatty acid methyl esters labelled with ^{14}C on diethylene glycol succinate, using proportional counter detection (20% CH_4, 80% Ar). A composition equivalent to that of inactive algae is found (16:0, 16:2, 18:1, 18:2 and 18:3 are the main components, see Fig. 3).

CONCLUSIONS

A gas chromatographic separation of the *Chlorella* higher fatty acids as their methyl esters is achieved only by chromatography on polar and non-polar stationary phases. In *Chlorella vulgaris* were found: (1) even-numbered straight-chain fatty acids as follows: saturated (C_{14}–C_{20}), mono-unsaturated (C_{14}–C_{18}), di-unsaturated (C_{14}–C_{18}), tri-unsaturated (C_{14}–C_{18}); (2) odd-numbered saturated fatty acids (C_{15}–C_{19}), both straight-chain and their *anteiso* derivatives.

ACKNOWLEDGEMENTS

This research was supported by the Czechoslovak Atomic Energy Commission, Prague. We thank Dr. J. HRADEC from the Oncological Institute for a kind gift of 14-methylhexadecanoic acid, Dr. J. DVOŘÁKOVÁ, ČSAV, Prague, for discussions of our results and Dr. J. KOLINA, ÚVVVR, Prague, for kind supply of cultivated *Chlorella vulgaris*.

REFERENCES

1 H. K. MANGOLD AND H. SCHLENK, *J. Biol. Chem.*, 229 (1957) 731.
2 H. SCHLENK, H. K. MANGOLD, J. L. GELLERMAN, W. E. LINK, R. A. MORRISSETTE, R. T. HOLMAN AND H. HAYES, *J. Amer. Oil Chem. Soc.*, 37 (1960) 547.

3 N. Pelick, R. S. Henly, R. F. Sweeny and M. Miller, *J. Amer. Oil Chem. Soc.*, 40 (1963) 419.
4 M. Karásek and J. Poláková, *Einige Erkenntnisse über die gas-chromatographische Trennung radiokohlenstoffmarkierter Fettsäuren*, Arbeitstagung Angewandte Radioaktivität, Leipzig, 1966.
5 R. F. Paschke and D. H. Wheeler, *J. Amer. Oil Chem. Soc.*, 31 (1954) 81.
6 A. T. James, *The Biosynthesis of Unsaturated Fatty Acids in Photosynthetic Systems in Lower and Higher Plants*, Proceedings of the 4th Meeting of FEBS, Oslo, 1967, Academic Press, New York, 1968.
7 R. V. Harris and A. T. James, *Biochim. Biophys. Acta*, 106 (1965) 465.
8 B. W. Nichols, A. T. James and J. Brener, *Biochem. J.*, 104 (1967) 486.
9 B. W. Nichols and A. T. James, in L. Reinhold and Y. Ch. Liwschitz (Editors), *Progress in Photochemistry*, Vol. I, Wiley, New York, 1968, p. 1.
10 E. J. Barron and L. A. Mooney, *Biochemistry*, 9 (1970) 2143.
11 S. Miyachi and S. Miyachi, *Plant Physiol.*, 41 (1966) 479.
12 A. G. Vereščagin and G. L. Kljačko-Gurvič, *Biokhimiya*, 30 (1965) 543.
13 I. Maslov, *Vestn. Leningr. Univ., Ser. Biol.*, 21 (1966) 102.
14 P. Pohl, H. Glasl and H. Wagner, *J. Chromatogr.*, 49 (1970) 488.
15 E. D. Korn, *J. Lipid Res.*, 5 (1964) 352.
16 Z. Nejedlý, J. Filip, J. Kolina, J. Ekl and D. Grünberger, *Czech. Pat.*, 121808, 1966.
17 M. Baslerová and J. Dvořáková, *Algarum, Hegaticarum Muscorumque in Culturic Collectio*, ČSAV, Prague, 1962.
18 J. Hradec, Oncological Institute, Prague, private communication.
19 J. Hradec, *J. Chromatogr.*, 32 (1968) 511.
20 J. Hradec and L. Dolejš, *Biochem. J.*, 107 (1968) 129.

Fatty Acid Synthetase from *Chlamydomonas reinhardi*

SITES OF TRANSCRIPTION AND TRANSLATION*

Reidun Sirevåg and Robert Paul Levine

SUMMARY

Cell-free extracts of the unicellular green alga, *Chlamydomonas reinhardi*, catalyze the incorporation of acetyl-CoA and malonyl-CoA into long chain fatty acids. The fatty acid synthetase is dependent on added acyl carrier protein for activity, regardless if the cells are grown in the light or in the dark. The major products formed from acetyl-CoA and malonyl-CoA by cells in the light period of synchronous growth are palmitate, stearate, and arachidate, whereas in the dark period of synchronous growth more short fatty acids are formed. Rifampicin has no effect on the appearance of the fatty acid synthetase in synchronous cultures of *C. reinhardi*, thus suggesting that transcriptional steps necessary for formation of fatty acid synthetase occur in the nucleus. Cycloheximide has no effect on the activity of fatty acid synthetase in the synchronous cultures, whereas spectinomycin reduced the activity significantly. These findings strongly indicate that the chloroplast ribosomes are responsible for the *de novo* protein synthesis required for fatty acid synthetase in *C. reinhardi*.

* This study was supported by Grants GB 1866 and GB 29203 from the National Science Foundation.

Current research has shown that although the chloroplast of the unicellular green alga *Chlamydomonas reinhardi* contains DNA and possesses the necessary equipment for protein synthesis (see Reference 1 for review), the synthesis of components of this organelle as well as its development are under control of both the nuclear-cytoplasmic genetic system and the genetic system of the chloroplast (2–4).

One characteristic feature of the chloroplast is its system of inner membranes or thylakoids that are known to be rich in galactolipids and sulfolipids (5). By their nature, the formation of these chloroplast-associated lipids has to be closely related to the biosynthesis of fatty acids, and thus when considering the genetic control over the formation of chloroplast components, it is of interest to determine the degree to which this organelle participates in the synthesis of fatty acids. In this paper we present the results obtained from an investigation of the synthesis of fatty acids and the cellular sites of transcription and translation of enzymes involved in their formation in *C. reinhardi*.

Antibiotics that selectively inhibit transcription and translation in either the nuclear-cytoplasmic system or in the chloroplast are important tools in studies on chloroplast formation. Rifampicin has been shown to inhibit RNA synthesis in the chloroplast of *C. reinhardi* (6, 7), *Chlorella pyrenoidosa* (8), and *Acetabularia* (9). Thus, the inhibitor apparently prevents the genes of the chloroplast from being transcribed. Spectinomycin has been shown to bind specifically to 70 S chloroplast ribosomes in *C. reinhardi* (10), while cycloheximide is known to inhibit protein synthesis on 80 S cytoplasmic ribosomes (2, 11). Thus, differential use of spectinomycin and cycloheximide gives information as to the sites of translation of various proteins. A system for synchronous growth of *C. reinhardi* has been described earlier (4) in which antibiotics were used in an attempt to determine the sites of genetic transcription and translation of certain important chloroplast components. Synchronous cultures are useful for such studies, since all the cells are in the same physiological state at any given time, and chloroplast and other cell components are synthesized at known times in preparation for a subsequent cellular division. Furthermore, an antibiotic can be added and its effects observed during a time period in which there is little RNA synthesis and the cells are nondividing. Also, the time of exposure of the cells to the antibiotic is relatively short, thus reducing the chances of secondary effects unrelated to protein synthesis.

Soluble enzyme systems that catalyze the formation of long chain fatty acids from acetyl-CoA and malonyl-CoA are known from a variety of organisms and tissues. In higher plants (12–14) and bacteria (15–17), the fatty acid synthetases consist of nonaggregating enzymes which can be separated into free

ACP[1] and enzyme components, which upon combination catalyze the synthesis of fatty acids, utilizing as substrates the ACP thioesters of the different acyl intermediates. The enzyme systems isolated from yeast (18) and animal tissue (19–22), on the other hand, are single multienzyme complexes, which are not dissociable into free ACP and active enzymes. Recently, it was reported that, depending on growth conditions, the phytoflagellate *Euglena gracilis* contains two independent fatty acids synthetases (23). When this organism was grown in the light on minimal medium, both an ACP-dependent and an ACP-independent fatty acid synthetase were present; the two different enzyme systems could be separated either by ammonium sulfate fractionation or by centrifugation in a sucrose density gradient. In etiolated cells of *E. gracilis*, however, only the ACP-independent fatty acid synthetase was present. By the differential use of cycloheximide and chloramphenicol, it was further shown in regreening cells of *E. gracilis* that *de novo* synthesis of the ACP-dependent fatty acid synthetase was completely dependent on the chloroplast ribosomes (24).

METHODS

Organism and Culture Conditions—*C. reinhardi*, a substrain of wild type strain 137c, mating type plus, was used. This strain is capable of synchronous, phototrophic growth in the minimal medium of Sueoka (25) as modified by Armstrong *et al.* (4). Synchronous growth was obtained by using a 12-hour light-12-hour dark cycle as described by Kates and Jones (26). Light and temperature conditions were as described by Surzycki (27). When rifampicin was used, the light intensity was increased from 5000 to 8000 lux, since this drug is light-absorbing. This was achieved by substituting tungsten reflector flood lamps for the white and daylight fluorescent lamps used for the synchronous cultures. The control flasks under these conditions were surrounded by a cell suspension that contained a rifampicin solution.

The experiments with synchronously grown cells were carried out during the third light period after inoculation, and at cell concentrations in the synchronous cultures of 0.5 to 1.0 $\times$ 10^6 cells per ml. The time of the onset of the third light period is designated zero time.

Mixotrophically grown cells, cells grown in the light on acetate medium, were obtained by growing the cultures in flasks with Tris-acetate-phosphate medium (28) at 25° on rotary shaker in the light (4000 lux from daylight fluorescent lamps). Hetero-

[1] The abbreviations used are: ACP, acyl carrier protein; DTT, dithiothreitol.

trophically grown cells, cells grown on acetate in the dark, were grown on the same medium in the dark.

Sterile conditions were maintained throughout.

Cell Number, Chlorophyll, and Protein—Cell number was determined with the aid of a hemacytometer on samples taken from the synchronous cultures at the beginning and end of the third light period. Chlorophyll was determined by a modification of the method (29) of Mackinney (30). Protein was measured by the Folin phenol reaction (31).

Inhibitors—Rifampicin (Mann Research Laboratories) was dissolved in sterile, unbuffered 0.01 M KH_2PO_4 to give a concentration of 15 mg per ml. It was added to the culture 50 min before the onset of the light period to give a final concentration of 250 μg per ml.

Spectinomycin (a gift from Upjohn Co.) was dissolved in distilled water to give a concentration of 5 mg per ml and sterilized by passing through a sterile Millipore membrane. The drug was added to the culture 50 min before the onset of the third light period to give a final concentration of 3 μg per ml unless otherwise stated.

Cycloheximide (Sigma Chemical Co.) was dissolved in distilled water to give a concentration of 1.5 mg per ml, sterilized by passing through a sterile Millipore membrane, and added to the cultures 3 hours after the onset of the third light period. The final concentration of cycloheximide in the cultures was 2 μg per ml.

Oxygen Evolution and Respiration—To measure oxygen evolution and respiration, a Yellow Springs Instrument Co. oxygen monitor (model 53) and a Clark-type electrode were used. A cell sample of 65 ml was taken from the synchronous culture, centrifuged at 1,000 $\times$ *g* for 5 min, and resuspended in Tris-acetate-phosphate medium containing 0.0025 M $NaHCO_3$. The measurements were done at 26° in a thermostated Lucite vessel. For photosynthetic oxygen evolution the light intensity was 20,000 lux.

Radioactive Fatty Acid Derivatives and ACP—[2-^{14}C]Malonyl-CoA and [^{3}H]acetyl-CoA, obtained from New England Nuclear, and *Escherichia coli* were very kindly provided by Professor K. Bloch.

Preparation of Enzyme—Samples of 400 ml were taken from the synchronous cultures and harvested by centrifugation at 17,000 $\times$ *g* for 10 min. The cells were washed once with and resuspended in 0.02 M potassium phosphate buffer (pH 7.0), 0.001 M in DTT to a final volume of 1.6 ml. Cell-free extracts were obtained by exposing the cells to sonic oscillation for two 1-min intervals with a Mullard 20-kc ultrasonic disintegrator operated at full power. The preparation was centrifuged for 10 min at 17,000 $\times$ *g* to get rid of cell debris. The resulting supernatant was centrifuged at 37,000 × *g* for 30 min. The soluble fraction (supernatant) was the crude extract.

Extracts of hetero- and mixotrophically grown cells were prepared in the same fashion, the concentration of cells before sonication being about 0.3 g per ml.

Ammonium sulfate fractionation of the crude extract was accomplished by adding a saturated salt solution to the crude extract to give 20, 35, 50, and 70% saturation. The precipitates were collected by centrifugation, dissolved to the original volume in the DTT-containing phosphate buffer, and assayed at once for enzyme activity. All operations were carried out at 0–4°.

Fatty Acid Synthetase Assay—Fatty acid synthetase activity was measured by the radioactive assay of Delo *et al.* (23). The reaction mixture routinely contained in a total volume of 0.5 ml: [2-^{14}C]malonyl-CoA, 30 nmoles; [^{3}H]acetyl-CoA, 15 nmoles; NADPH, 15 nmoles; NADH, 15 nmoles; FMN, 0.5 nmole; DTT, 2.5 μmoles, potassium phosphate buffer (pH 7.2), 0.5 mmole; *E. coli* ACP, 10 μmoles; and 0.1 ml of enzyme preparation. The assay mixtures were incubated for 15 min at 30°, the reaction stopped by adding 0.15 ml of 50% KOH, and the suspension saponified at 100° for 30 min. After acidification with 8 to 9 drops 6 N HCl, the fatty acids were extracted three times with 5 ml of petroleum ether (b.p. 37–46.2°). The solvent was evaporated and the residue dissolved in 15 ml of 2,5-diphenyloxazole (PPO)-toluene and counted in a Packard Tri-Carb liquid scintillation spectrophotometer, model 574.

The incorporation of [2-^{14}C]malonyl-CoA and [^{3}H]acetyl-CoA into long chain fatty acids was determined by computer analyses of the counting data. In experiments with synchronously grown cells, the results were calculated on the basis of 100 ml of synchronous culture, containing 10^6 cells per ml, since the cell number remains constant in these cultures during the experimental period.

For heterotrophically grown cells, the results are expressed on the basis of cell wet weight.

When both [^{3}H]acetyl-CoA and [2-^{14}C]malonyl-CoA are used as substrates, the two isotopes will be present in the reaction products. Thus, the ratio of ^{14}C:^{3}H incorporated into long chain fatty acids permits an estimation of the chain length of these acids (24).

Identification of Fatty Acids—The petroleum ether extracts containing total fatty acids were taken to dryness in a stream of nitrogen. The residues were esterified by heating with 0.2 ml of 14% BF_3 in methanol at 60° for 3 min, neutralized with NaOH, and extracted three times with petroleum ether. The solvent was removed under nitrogen and the methyl esters purified by passing through a silicic acid column (Unicil, Clarkson Chemical Co., Williamsport, Pa.), which was eluted with 5%

TABLE I

Dependence on ACP of fatty acid synthetase from C. reinhardi, grown phototrophically in synchronous cultures

Cells were grown synchronously as described under "Methods" and harvested after 6 hours growth in the third light period. Crude extracts and $(NH_4)_2SO_4$ fractions were prepared and assayed for fatty acid synthetase activity as described. The reaction mixture for the radioactive assay contained in a total volume of 0.5 ml: [2-^{14}C]malonyl-CoA, 30 nmoles; [^{3}H]acetyl-CoA, 15 nmoles; NADPH, 15 nmoles; NADH, 15 nmoles; FMN, 0.5 nmole; DTT, 2.5 μmoles; potassium phosphate buffer (pH 7.2), 0.5 mmole; *E. coli* ACP, 10 μmoles; and 0.1 ml of enzyme preparation. Time of incubation: 15 min at 30°. The values given are nanomoles of [2-^{14}C]malonyl-CoA plus [^{3}H]acetyl-CoA incorporated into fatty acids by 100 ml of synchronous culture containing 10^6 cells per ml.

Source of enzyme	Amount [2-^{14}C]malonyl-CoA plus [^{3}H]acetyl-CoA incorporated into fatty acids	
	With ACP	Without ACP
	nmoles	
Crude extract	1.100	0.061
Fraction 0 to 20% $(NH_4)_2SO_4$	0.008	0.000
Fraction 20 to 35% $(NH_4)_2SO_4$	0.003	0.000
Fraction 35 to 50% $(NH_4)_2SO_4$	0.004	0.000
Fraction 50 to 70% $(NH_4)_2SO_4$	0.825	0.008

diethyl ether in petroleum ether.

Radioactive fatty acid methyl esters were separated by gas chromatography together with methyl esters of carrier fatty acids (C_{10}, C_{12}, C_{14}, C_{16}, C_{18}, C_{20}, and oleic acid). All analyses were carried out on a 6-foot column of 10% diethylene glycol succinate in an F and M model 400 instrument equipped with a hydrogen flame detector and variable beam splitter. The column temperature was programmed for an increase of 4° per min to 170°. The effluent was passed directly to a proportional counter (Nuclear-Chicago).

To estimate the ratio of one long chain fatty acid formed to another, the individual radioactive peaks displayed by the tracings from gas chromatography were cut out and weighed. The percentage of each peak was calculated on the basis of the total weight of all radioactive peaks.

RESULTS

Fatty Acid Synthetase from Synchronously Grown Cells of C. reinhardi—Crude extracts from synchronously grown photo-

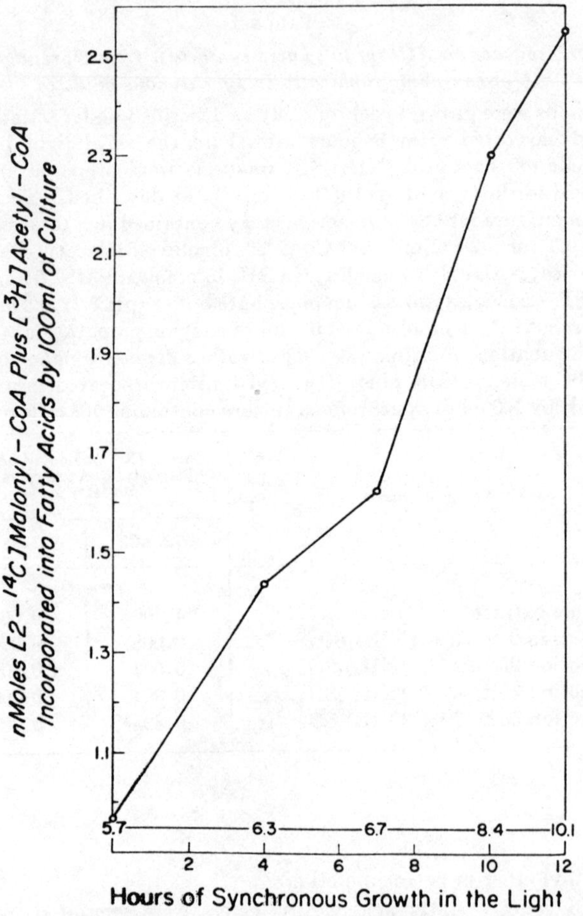

FIG. 1. The increase in ACP-dependent fatty acid synthetase during the third light period of synchronous growth. Cells were harvested, broken, and assayed for the enzyme as described under "Methods" and Table I. The numbers inserted above the abscissa represent the numerical values of the ^{14}C:^{3}H ratio at different times in the light period.

trophic cells of *C. reinhardi* incorporated malonyl-CoA and acetyl-CoA into fatty acids in the presence of the cofactors described under "Methods." The incorporation was greatly stimulated upon addition of *E. coli* ACP (Table I). Fractionation of the crude extract showed that the enzyme activity was recovered in the protein precipitating between 50 and 70% saturated ammonium sulfate (Table I), as is the case for the ACP-dependent enzyme in *E. gracilis* (23).

During the third light period of synchronous growth of *C. reinhardi*, the activity of the fatty acid synthetase increases from

TABLE II

Saturated fatty acids produced by C. reinhardi fatty acid synthetase at different times in third light period of synchronous growth

Cells were grown synchronously, and samples taken at times indicated. Extracts were prepared and assayed for fatty acid synthetase activity in the presence of *E. coli* ACP as described under "Methods" and Table I. Individual fatty acids were analyzed by separation of their methyl esters by gas chromatography as described. The results are expressed as a percentage of the total fatty acids formed by crude extracts.

Time of synchronous growth in the light	Fatty acid produced, percentage of total				
	C_{12}	C_{14}	C_{16}	C_{18}	C_{20}
hrs	%	%	%	%	%
0	14.8	20.9	33.0	31.4	0
6	0	0	19.8	66.5	13.7
12	0	0	18.3	64.7	17.0

TABLE III

Dependence on ACP of fatty acid synthetase from mixotrophically grown C. reinhardi

The cells were grown mixotrophically for 40 hours as described under "Methods," harvested, washed, and resuspended in potassium phosphate buffer (pH 7.0), 0.001 M in DTT, to give a final concentration of 0.3 mg per ml, wet weight, before sonication. Crude extracts and ammonium sulfate fractions were assayed for fatty acid synthetase activity as described under "Methods" and Table I, except that unlabeled acetyl-CoA was used. The results are expressed as nanomoles [2-^{14}C]malonyl-CoA incorporated into fatty acids by an amount of cell-free extract corresponding to 1 g of cells, wet weight.

Source of enzyme	Amount protein per g cells, wet weight	Amount of [2-^{14}C]malonyl-CoA incorporated into fatty acids	
		With ACP	Without ACP
	mg	*nmoles*	
Crude extract	18.5	8.23	0.21
Fraction 0 to 20% $(NH_4)_2SO_4$	2.4	0.02	0.01
Fraction 20 to 35% $(NH_4)_2SO_4$	8.3	0.20	0.01
Fraction 35 to 50% $(NH_4)_2SO_4$	5.6	0.03	0.59
Fraction 50 to 70% $(NH_4)_2SO_4$	3.2	12.07	0.04

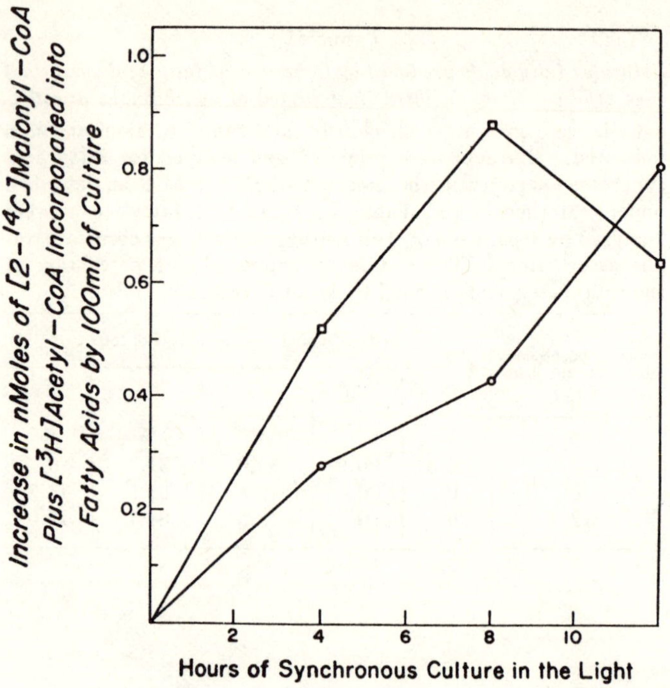

FIG. 2. Effect of rifampicin on the increase in ACP-dependent fatty acid synthetase during the third light period of synchronous growth. Cells were harvested, broken, and assayed for the enzyme as described under "Methods" and Table I. ○, control; □, 250 μg per ml of rifampicin. Rifampicin was added 50 min before zero time.

the 1st to the 12th hour. The level of activity varied somewhat from one experiment to another, but in general the activity of the synthetase increased in the fashion shown in Fig. 1.

The major products formed from malonyl-CoA and acetyl-CoA by the fatty acid synthetase from extracts of synchronously grown *C. reinhardi* were identified by gas chromatography as palmitate (C_{16}), stearate (C_{18}), and arachidate (C_{20}). At the onset of the light period some C_{12} and C_{14} fatty acids were present in the reaction product, while C_{20} fatty acids were not found at this time. The ratio of one long chain fatty acid formed to another at different times in the third light period of synchronous growth is shown in Table II. As can be seen the formation of the longer (C_{20}) fatty acids increases with time, and apparently at the expense of the shorter fatty acids. This increase in formation of longer chain fatty acids with time also was confirmed on the basis of ^{14}C:^{3}H ratios, which represent the relative incorporation of malonyl-CoA to acetyl-CoA into fatty acids (24). Thus, at the onset of the third light period, this ratio had a numerical

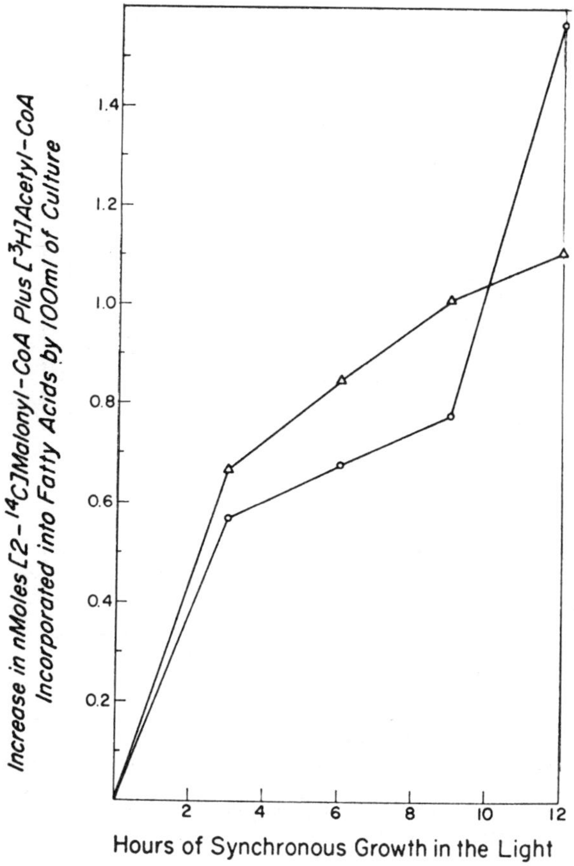

FIG. 3. Effect of cycloheximide on the increase in ACP-dependent fatty acid synthetase during the third light period of synchronous growth. Cells were harvested, broken, and assayed for the enzyme as described under "Methods" and Table I. ○, control; △, 2 μg per ml of cycloheximide. Cycloheximide was added 3 hours after zero time.

value of about six, which corresponds to an average length of 14 carbon atoms in the fatty acids formed. The $^{14}C:^{3}H$ ratio at the end of the light period had a value of 8 to 9, corresponding to an average of 18 carbon atoms in the fatty acid chain.

Fatty Acid Synthetase from Cells of C. reinhardi Grown Mixotrophically and Heterotrophically—When cell-free extracts prepared from mixotrophically grown cells of *C. reinhardi* were assayed for fatty acid synthetase, it was found that addition of *E. coli* ACP stimulated the incorporation of malonyl-CoA and acetyl-CoA into fatty acids in the same manner as for the enzyme

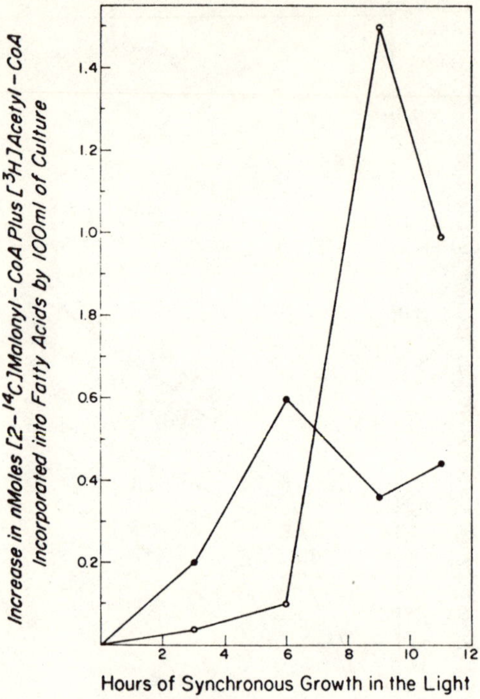

FIG. 4. Effect of spectinomycin on the increase in ACP-dependent fatty acid synthetase during the third light period of synchronous growth. Cells were harvested, broken, and assayed for the enzyme as described under "Methods" and Table I. ○, control; ●, 3 μg per ml of spectinomycin. Spectinomycin was added 50 min before zero time.

from phototrophically grown cells. Further, as was the case for the enzyme from phototrophically grown cells, ammonium sulfate precipitated the enzyme between 50 and 70% saturation. This is shown in Table III. The enzyme from heterotrophically grown cells showed the same pattern.

Effect of Inhibitors on Synthesis of Fatty Acid Synthetase in Synchronous Cultures of C. reinhardi—Rifampicin did not affect the increase in fatty acid synthetase during the third light period of synchronous growth as is shown in Fig. 2. The relatively low values for enzyme activity obtained for both the treated and the control are probably due to the light conditions that prevail in experiments with rifampicin (see "Methods").

The synthesis of the ACP-dependent fatty acid synthetase in cell-free extracts of *C. reinhardi* was not affected by cycloheximide (Fig. 3), whereas chlorophyll synthesis in this experiment was inhibited by about 90%.

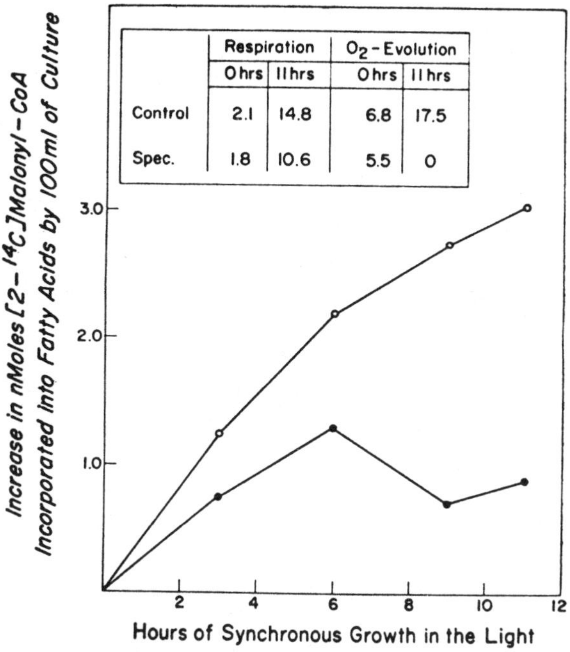

FIG. 5. Effect of an increased concentration of spectinomycin on the increase in ACP-dependent fatty acid synthetase. Cells were grown, harvested, broken, and assayed for the enzyme as described under "Methods" and Table I, except that unlabeled acetyl-CoA was used. The *inserted table* shows the effect of spectinomycin on the increase in respiration and photosynthetic oxygen evolution. Samples were taken at zero time and after 11 hours and respiration and photosynthetic oxygen evolution measured as described under "Methods". The results are expressed as micromoles of O_2 consumed or evolved per hour by 65 ml of synchronous cultures. ○, control; ●, 4.5 μg per ml of spectinomycin. Spectinomycin was added 50 min before zero time.

When spectinomycin was added to the synchronously growing cells, the synthesis of the ACP-dependent fatty acid synthetase increased in the first 6 hours after the onset of the light period, after this period the synthesis went down. This was the common pattern when cells were exposed to this drug. Results from a typical experiment are shown in Fig. 4. Although the rate of synthesis of the enzyme in treated cultures always begins to go down around the 6th hour, the early (0 to 3 hours) rapid rate in treated cultures is observed to vary from one experiment to another (compare Figs. 4 and 5). At the end of the light period the cells were viable and the level of chlorophyll was the same as for cells in the control.

To test the possibility that the decrease in appearance of the enzyme was due to secondary effects of the drug, *i.e.* inhibition of respiration, an experiment was performed in which the cells were exposed to higher concentrations of spectinomycin (4.5 μg per ml in contrast to 3 μg per ml usually employed). The rates of respiration and photosynthetic oxygen evolution by whole cells were measured at the onset of the third light period and after 11 hours in the light. The data from this experiment are given in Fig. 5. As is shown in the *inset* of the figure, photosynthetic oxygen evolution was completely inhibited by spectinomycin, whereas the increase in respiratory capacity of the cells was only slightly affected (4). The appearance of the fatty acid synthetase showed the same pattern as in the experiment shown in Fig. 4.

DISCUSSION

The chloroplast of *C. reinhardi* has a certain degree of genetic autonomy in that it contains DNA and a complete protein-synthesizing machinery (1). However, so far only a few components of the chloroplast have been found that are coded for by the chloroplast DNA (4). In the experiments where the synchronously grown cells of *C. reinhardi* were exposed to rifampicin, an inhibitor of algal chloroplast RNA synthesis (6–9), the activity of the fatty acid synthetase increased in the same fashion as in the control cultures, thus suggesting that the genetic information necessary for synthesis of this enzyme system is located in the nucleus. It should be kept in mind, however, that a lack of an effect of rifampicin must be interpreted cautiously since the lifetime of chloroplast messenger RNA is not known. Although the chloroplast of *C. reinhardi* may not be involved in transcriptional steps that lead to the formation of the fatty acid synthetase, the synthesis of the enzyme depends on translational steps that occur on chloroplast ribosomes. Spectinomycin, which binds specifically to the 70 S chloroplast ribosomes in *C. reinhardi* (10) inhibits the formation of the fatty acid synthetase drastically, under conditions where chlorophyll synthesis, which depends on translational steps on cytoplasmic ribosomes (4), was not affected at all. Furthermore, when the synchronous cultures were treated with cycloheximide, the inhibitor of protein synthesis on the 80 S cytoplasmic ribosomes, at a concentration (2 μg per ml) which inhibited chlorophyll synthesis by 90%, they contained the same level of fatty acid synthetase as did the control cultures. Thus, the results obtained by the differential use of spectinomycin and cycloheximide strongly indicate that *de novo* protein synthesis necessary for synthesis of fatty acids take place on the 70 S ribosomes of the chloroplast in *C. reinhardi*.

It appears from the results reported here that *C. reinhardi* contains a fatty acid synthetase similar to the one reported to occur in green plants (12–14) and bacteria (17). The fatty acid synthetase from *C. reinhardi* is completely dependent on the presence of ACP, regardless whether the organism is grown in the light or in the dark, and the major products formed from acetyl-CoA and malonyl-CoA in the light are palmitic, stearic, and arachidic acids. In contrast, it has been reported that *E. gracilis* has the ability to form two distinct and separable fatty acid synthetases depending on the growth conditions. The two fatty acid synthetases from *E. gracilis* differ from each other in molecular properties and give rise to fatty acids with different chain lengths (23). Thus *C. reinhardi* differs from *E. gracilis* with respect to the synthesis of fatty acids. This difference between *C. reinhardi* and *E. gracilis* in their enzymes for fatty acid synthesis probably mirrors the difference in mode of life of these two types of organisms. In *C. reinhardi*, both chlorophyll synthesis and chloroplast formation are constitutive processes which occur in the dark as well as the light (32), whereas in *E. gracilis* these processes are absent in the dark, and are only induced when the cells are exposed to the light (33).

Although only one class of fatty acid synthetase was found in cell-free extracts of *C. reinhardi*, it appears that the types and amounts of fatty acids formed are somewhat dependent on the stage of the cell cycle. This is indicated by the results from experiments with synchronous cultures, where the fatty acids formed in the dark (zero time) had shorter chain lengths than those formed later in the third light period (Table II and Fig. 1). When the cells in synchronous culture enter into a light period, many of their chloroplast components start to increase in amount in preparation for the cell division that will occur in the subsequent dark period. Thus the fact that relatively more of the long chain fatty acids are produced later in the light period probably reflects the need of the cell for precursors for oleic and α-linolenic acids, the major unsaturated fatty acids in the chloroplast membranes of *C. reinhardi* (23).

Acknowledgments—We wish to thank Professor Konrad Bloch for his generosity in providing the radioactive fatty acid derivatives and the ACP necessary for the experiments described here, for his advice and suggestions, and for the use of facilities in his laboratory. We also thank Dr. Barbara Talamo and Dr. Harold Wright for their assistance during the course of the experiments.

REFERENCES

1. Levine, R. P., and Goodenough, U. W. (1970) *Annu. Rev. Genet.* **4**, 397

2. Hoober, J. K., Siekevitz, P., and Palade, G. E. (1969) *J. Biol. Chem.* **244**, 2621-2631
3. Hoober, J. K. (1970) *J. Biol. Chem.* **245**, 4327–4334
4. Armstrong, J. J., Surzycki, S. J., Moll, B., and Levine, R. P. (1971) *Biochemistry* **10**, 692
5. Benson, A. A. (1964) *Annu. Rev. Plant Physiol.* **15**, 1
6. Surzycki, S. J. (1969) *Proc. Nat. Acad. Sci. U.S.A.* **63**, 1327
7. Surzycki, S. J., and Rochaix, J. (1971) *J. Mol. Biol.* **62**, 89
8. Galling, G. (1971) *Planta* **98**, 50
9. Brandle, E., and Zetsche, K. (1971) *Planta* **99**, 46
10. Burton, W. G. (1970) *J. Cell Biol.* **47**, 28a
11. Siegal, M. R., and Sisler, H. D. (1964) *Biochim. Biophys. Acta* **87**, 83–89
12. Overath, P., and Stumpf, P. K. (1964) *J. Biol. Chem.* **239**, 4103–4110
13. Brooks, J. L., and Stumpf, P. K. (1966) *Arch. Biochem. Biophys.* **116**, 108
14. Simoni, R. D., Criddle, R. S., and Stumpf, P. K. (1967) *J. Biol. Chem.* **242**, 573–581
15. Lennarz, W. J., Light, R. J., and Bloch, K. (1962) *Proc. Nat. Acad. Sci. U.S.A.* **48**, 840
16. Alberts, A. W., Goldman, P., and Vagelos, P. R. (1963) *J. Biol. Chem.* **238**, 557–565
17. Goldman, P., Alberts, A. W., and Vagelos, P. R. (1963) *J. Biol. Chem.* **238**, 1255–1261
18. Lynen, F. (1961) *Fed. Proc.* **20**, 941
19. Bressler, R., and Wakil, S. J. (1961) *J. Biol. Chem.* **236**, 1643–1651
20. Butterworth, P. H. W., Yang, P. C., Bock, R. M., and Porter, J. W. (1967) *J. Biol. Chem.* **242**, 3508–3516
21. Erwin, J., and Bloch, K. (1962) *Biochem. Biophys. Res. Commun.* **9**, 103
22. Burton, D., Haarik, A. G., and Porter, J. W. (1968) *Arch. Biochem. Biophys.* **126**, 141
23. Delo, J., Ernst-Fonberg, M. L., and Bloch, K. (1971) *Arch. Biochem. Biophys.* **143**, 384
24. Ernst-Fonberg, M. L., and Bloch, K. (1971) *Arch. Biochem. Biophys.* **143**, 392
25. Sueoka, N. (1960) *Proc. Nat. Sci. U. S. A.* **46**, 83
26. Kates, J. R., and Jones, R. F. (1964) *J. Cell Comp. Physiol.* **63**, 157
27. Surzycki, S. J. (1971) *Methods Enzymol.* **23**, 67–73
28. Gorman, D. S., and Levine, R. P. (1965) *Proc. Nat. Acad. Sci. U. S. A.* **54**, 1665
29. Arnon, D. I. (1949) *Plant Physiol.* **24**, 1
30. Mackinney, G. (1941) *J. Biol. Chem.* **140**, 315
31. Lowry, O. H., Rosebrough, N. J., Farr, A. L., and Randall, R. J. (1951) *J. Biol. Chem.* **193**, 265–275
32. Levine, R. P., and Ebersold, W. T. (1960) *Annu. Rev. Microbiol.* **14**, 197
33. Schiff, J. A., and Epstein, H. T. (1968) in D. E. Buetow (Editor), *The biology of euglena, Vol. II*, p. 285, Academic Press New York

Effects of Pesticides

Polychlorinated Biphenyls and DDT Alter Species Composition in Mixed Cultures of Algae

JERRY L. MOSSER
NICHOLAS S. FISHER
CHARLES F. WURSTER

Abstract. *Either DDT or polychlorinated biphenyls were added to mixed cultures containing a marine diatom and a marine green alga that were sensitive and resistant, respectively, to these organochlorine compounds. The diatom grew faster and was therefore dominant in control cultures, but its dominance diminished in treated cultures, even at concentrations of chlorinated hydrocarbons that had no apparent effect in pure cultures. Such stable pollutants could disrupt the species composition of phytoplankton communities, thereby affecting whole ecosystems.*

The impact of certain persistent chlorinated hydrocarbons on various higher nontarget organisms has been well documented (*1*), but effects on photosynthetic algae, the base of aquatic food webs, have not been extensively studied. Marine phytoplankton vary in sensitivity to chlorinated hydrocarbons, including DDT [1,1,1-trichloro-2,2-bis (*p*-chlorophenyl)ethane] and polychlorinated biphenyls (PCBs). Some species show effects at concentrations as low as a few parts per billion (ppb), whereas others are resistant to much higher concentrations (*2, 3*). Some of these chemicals, especially DDT and PCBs, are extremely widespread pollutants of the biosphere, and, because they are selectively toxic to certain sensitive algal species, it has been hypothesized that they could alter the species composition of phytoplankton communities (*2, 3*). Evidence for this hypothesis is lacking; we therefore investigated the effects of DDT and PCBs in mixed algal cultures containing a sensitive and a resistant species.

Two marine organisms were selected on the basis of their sensitivity to organochlorine compounds: growth of the diatom *Thalassiosira pseudonana* was inhibited by PCBs and DDT, whereas *Dunaliella tertiolecta*, a green alga, was not affected by these chemicals (*3*). Methods of culture and procedures for treatment have been described (*3*). Cultures contained a total of 10^4 exponentially growing cells per milliliter at zero time; mixed cultures contained the two species in a 1:1 ratio. Mixed and pure cultures were treated simultaneously. Cells in pure culture were counted with a Coulter counter, whereas those in mixed cultures were counted microscopically with a Neubauer-Levy counting chamber because the two cell types could not be differentiated by the Coulter counter.

The results with pure cultures confirmed those described previously (*3*). Polychlorinated biphenyls inhibited the growth of *T. pseudonana* at 25 ppb, but not at 10 ppb or less. Growth was also significantly inhibited by DDT at 100 ppb; inhibition at 50 ppb was slight, and none occurred at 25 ppb or lower. By contrast, the growth of *D. tertiolecta* was not affected at any of the concentrations of PCBs or DDT tested.

In untreated mixed cultures, *T. pseudonana* reproduced more rapidly than *D. tertiolecta*, reaching an eight- to ninefold greater cell concentration after 4 days. Treatment with PCBs or DDT significantly diminished the competitive success of *T. pseudonana* and increased that of *D. tertiolecta* at all concentrations tested, even though the lowest concentrations (1 and 10 ppb, respectively) had no detectable effect on the growth of *T. pseudonana* in pure cultures (Fig. 1, A and B). The graphs of species ratios (Fig. 1C) suggest that still lower concentrations would alter the final species composition of the cultures. Higher concentrations of the organochlorine compounds caused greater deviation in species ratios from those of the control cultures.

Although species ratios were substantially changed, the final total cell numbers did not differ significantly among control and treated mixed cultures. Since cells of the two species are approximately the same size, the final biomass was not markedly affected. This result suggests that the two organisms were competing for a limiting nutrient in the mixed cultures. Presumably *T. pseudonana* assimilated most of the nutrient in control cultures because it grew faster, but its ability to compete was impaired by PCBs and DDT; more nutrient then became available for *D. tertiolecta*. Since phytoplankters often compete for limiting resources in nature (*4*), tests evaluating pollutants in mixed cultures probably give a more ecologically meaningful indication of algal sensitivity than those in pure cultures. *Thalassiosira pseudonana* was substantially more sensitive to PCBs and DDT in mixed cultures than in pure cultures.

Chlorinated hydrocarbons and other stable pollutants (*5*) could alter natural algal communities by suppressing sensitive species and permitting pollutant-resistant forms to become dominant. Polychlorinated biphenyls and DDT can occur in natural waters at concentrations comparable to those that altered the species ratios in our experiments, although concentrations found in nature are generally lower (*6*). Exposure to organochlorine compounds is probably greater than the concentrations in natural waters would indicate, however, because these substances are rapidly absorbed from water by organisms, including phytoplankters (*1*). In eutrophic environments, alterations of algal communities could further reduce an already decreased species diversity (*7*), aggravating problems of algal blooms and contributing to the general degradation of the ecosystem (*8*).

Many zooplankters graze selectively, often choosing their food on the basis of size or shape (*9, 10*). The dietary requirements of herbivores are not satisfied by all algal species, as indicated by the growth rates of oyster and clam larvae (*11*), the viability of

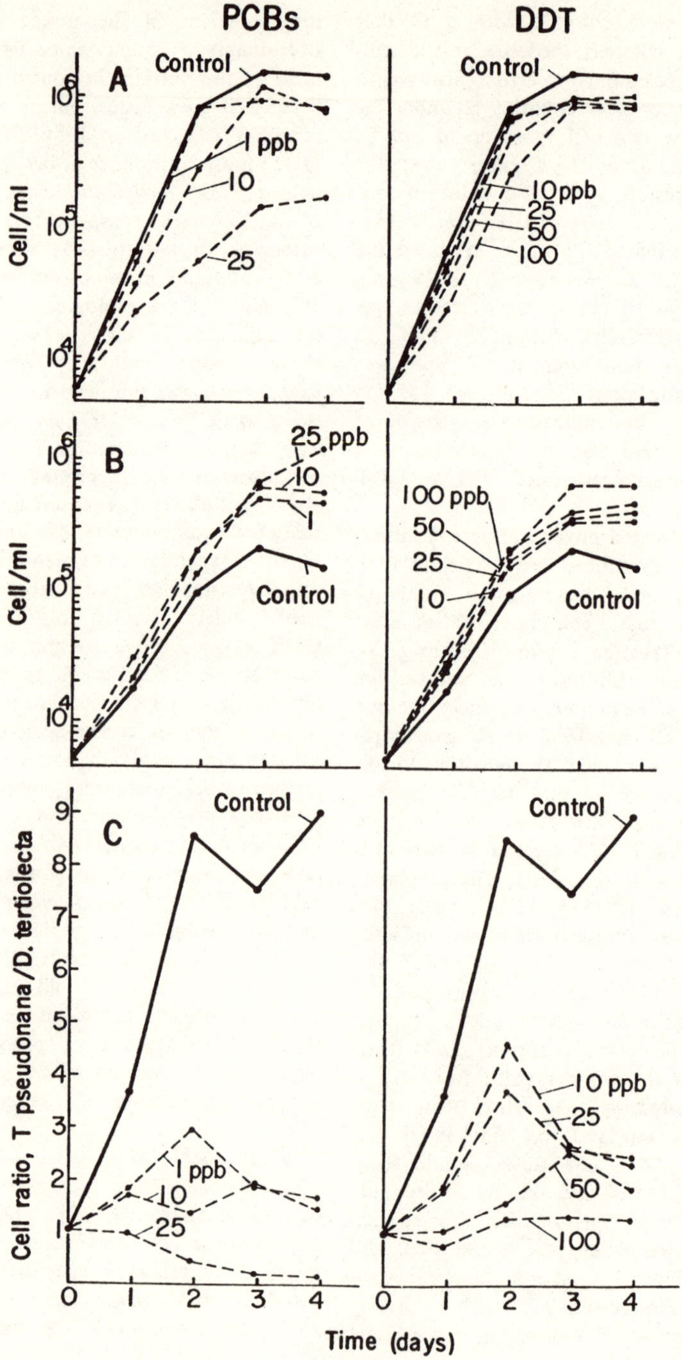
PCBs
DDT
A
B
C
Control
1 ppb
10
25
10 ppb
50
100
25 ppb
1
100 ppb
Cell/ml
Cell ratio, T pseudonana/D. tertiolecta
Time (days)

Fig. 1. Growth of (A) *T. pseudonana* and (B) *D. tertiolecta* in mixed cultures; (C) species ratios. Data points are the means of three replicates. Cell concentrations: single classification analyses of variance showed that, at day 4, *D. tertiolecta* controls differed from all treated cultures ($P < .005$); *T. pseudonana* controls differed from all treated cultures ($P < .025$), except at 25 ppb of DDT ($P < .10$). Species ratios: analyses of variance, performed by means of the Kruskal-Wallis and Wilcoxon two-sample tests (*13*) revealed that, on days 3 and 4, controls differed from all treated cultures ($P < .001$); ratios for 100 ppb of DDT differed from those for 50 ppb of DDT, and ratios for 25 ppb of PCBs differed from those for 10 ppb of PCBs ($P < .05$).

barnacle nauplii (*10*), and reproduction in copepods (*12*). Hence, altering the species composition of a phytoplankton community could profoundly affect the health, distribution, and abundance of many animal populations higher in the food web (*8, 10, 11*).

References and Notes

1. D. B. Peakall, *Science* **168**, 592 (1970); *Sci. Amer.* **222** (4), 72 (1970); ——— and J. L. Lincer, *BioScience* **20**, 958 (1970); K. J. Macek, *J. Fish. Res. Board Can.* **25**, 1787 (1968); *ibid.*, p. 2443; C. F. Wurster, *Environment* **13** (8), 33 (1971); *Biol. Conserv.* **1**, 123 (1969); *Proceedings of the International Conference on the Environment of the Future*, Jyvaskyla, Finland, 1 July 1971 (Macmillan, London, in press) (*Congr. Rec.*, 28 July 1971, p. E8333).
2. C. F. Wurster, Jr., *Science* **159**, 1474 (1968); D. W. Menzel, J. Anderson, A. Randtke, *ibid.* **167**, 1724 (1970).
3. J. L. Mosser, N. S. Fisher, T.-C. Teng, C. F. Wurster, *ibid.* **175**, 191 (1972).
4. G E. Hutchinson, *A Treatise on Limnology* (Wiley, New York, 1967), vol. 2, pp. 355–489; E. M. Hulburt, *Ecology* **51**, 475 (1970).
5. R. C. Harriss, D. B. White, R. B. Macfarlane, *Science* **170**, 736 (1970).
6. T. W. Duke, J. I. Lowe, A. J. Wilson, *Bull. Environ. Contam. Toxicol.* **5**, 171 (1970); J. O. Keith and E. G. Hunt, *Trans. North Amer. Wildlife Natur. Resour. Conf., 31st*, 14–16 March 1966 (1966), pp. 150–77.
7. R. Patrick, *Ann. N.Y. Acad. Sci.* **108**, 359 (1963); C. M. Palmer, *ibid.*, p. 389; R. Margalef, *Oceanogr. Mar. Biol. Annu. Rev.* **5**, 257 (1967).
8. G. M. Woodwell, *Science* **168**, 429 (1970); J. H. Ryther, *Biol. Bull.* **106**, 198 (1954).
9. V. Bainbridge, *J. Mar. Biol. Ass. U.K.* **37**, 349 (1958); H. J. Curl and G. C. McLeod, *J. Mar. Res.* **19**, 70 (1961); J. J. Lee, M. McEnery, S. Pierce, H. D. Freudenthal, W. A. Muller, *J. Protozool.* **13**, 659 (1966); S. Richman and J. N. Rogers, *Limnol. Oceanogr.* **14**, 701 (1969).
10. E. J. F. Wood, *Marine Microbial Ecology* (Reinhold, New York, 1965), pp. 71–96.
11. H. C. Davis and R. R. Guillard, *U.S. Fish Wildlife Serv. Fish. Bull. 136* **58**, 293 (1958).
12. S. M. Marshall and A. P. Orr, *J. Mar. Biol. Ass. U.K.* **30**, 527 (1952); L. Provasoli, K. Shiraishi, J. R. Lance, *Ann. N.Y. Acad. Sci.* **77**, 250 (1959).
13. R. R. Sokal and F. J. Rohlf, *Biometry* (Freeman, San Francisco, 1969), pp. 388–94.

14. We thank M. F. Luhnow for laboratory assistance and R. R. L. Guillard for valuable consultations. Supported by NSF grant GB-11902 and NIH postdoctoral fellowship 1FO2ES48112-01 to J.L.M.

3-(3,4-DICHLOROPHENYL)-1,1-DIMETHYLUREA (DCMU) INHIBITION OF SYSTEM II AND LIGHT-INDUCED REGULATORY CHANGES IN ENERGY TRANSFER EFFICIENCY

L. N. M. DUYSENS

ABSTRACT Oxygen pulses produced in *Chlorella* by a xenon flash of 15 μsec half-width were measured by means of a rapid oxygen polarograph. Under appropriate conditions the height of the pulse caused by a saturating flash was a measure of the number of active reaction centers in system II. In pigment state II, caused by illumination during several minutes with light II, the number of active centers II was the same as in pigment state I. Oxygen pulses produced by about half-saturating flashes were diminished by about 7–10% in state II, showing that the fluorescence decrease in light II was at least partly caused by a decrease in energy transfer to reaction center II. After addition of 3(3,4-dichlorophenyl)-1,1-dimethylurea (DCMU), only the first flash produced oxygen which gives additional support for the hypothesis that DCMU inhibits between Q and sytem I.

INTRODUCTION

The experiments described in this paper will be discussed in terms of the following model (Duysens and Sweers, 1963). Light energy absorbed by or transferred to chlorophyll a_2 molecules (of photosystem II) is transferred via a number of chlorophyll a_2 molecules to a specialized chlorophyll a molecule C in a reaction center. The molecule C forms a complex ZCQ with an electron donor Z and an electron acceptor Q. The absorption and fluorescence spectra of C do not differ too much from those of the light-harvesting chlorophyll a_2. Only the fluorescence yield of the complex ZCQ must be much lower than that of the light-harvesting chlorophyll a in order to make efficient trapping possible.

The excitation energy is rapidly used for the transfer of an electron from Z to Q:

$$ZCQ + h\nu_2 \rightarrow ZC^*Q \rightarrow Z^+CQ^-. \qquad (1)$$

The ratio of the number of chlorophyll a_2 molecules to that of reaction centers II is of the order of 200, which is of the same order as the number of steps taken to reach the reaction center (Duysens, 1964; Knox, 1968).

Since the chlorophyll a_2 fluorescence increases upon reduction of Q by illumination with light II, or by addition of the reductant dithionite, and decreases upon oxidation by light I, it has been concluded (Duysens and Sweers, 1963) that $Z^{(+)}C^*Q^-$ has a relatively long lifetime and is thus able to transfer the excitation back to the bulk of chlorophyll a_2 which causes an appreciable increase in fluorescence yield. It should be kept in mind that whatever the state of the complex, at least 99.5% of the fluorescence is emitted by the light-harvesting chlorophyll, and less than $1/200$ by the reaction center.

Before there was good evidence about the existence of such complexes Rabinowitch (1951) discussed in a general way possible properties of the various forms of the complexes, especially the fluorescence yields. He pointed out that nothing could be said a priori about the fluorescence yield of the forms Z^+CQ, ZCQ^-, and Z^+CQ^-, since the oxidized and reduced forms of Z and Q may quench or "dequench" the fluorescence yield of C (e.g., by speeding up or slowing down internal conversion). Furthermore the fluorescence yield of the light-harvesting chlorophyll may be influenced by conformational and other changes in the thylakoid membrane in which the pigments are imbedded.

It should be kept in mind that such possible complications make part of the conclusions in this paper somewhat uncertain; however, we will discuss the experiments in terms of one model which appears most plausible, and which can be extended, modified, or rejected by further evidence.

Even if the primary reaction would be, which seems plausible, a rapid electron transfer

$$ZC^*Q \rightarrow ZC^+Q^-, \tag{2}$$

followed by

$$ZC^+Q^- \rightarrow Z^+CQ^-, \tag{3}$$

we would not be able to state for certain that the fluorescence yield of Z^+CQ is small. The reason is that the charge on Z may change the conformation of the complex so that the reaction $Z^+CQ + h\nu_2 \rightarrow Z^+C^+Q^-$ cannot occur. Thus the fluorescence yield of Z^+CQ might be higher than that of ZCQ. The experiments reported in this paper have been done to check and possibly extend conclusions based on the model discussed.

METHODS

A thin layer of *Chlorella* cells was deposited on a stationary horizontal platinum electrode. The other electrode was a cylindrical silver one surrounding the platinum electrode. The voltage between the electrodes was 0.7 v. The algae were covered by a GE "Nucleopore" membrane with holes of 0.5 μ. This membrane permits rapid diffusion of inhibitors in contrast to a cellophane membrane. The cells were illuminated with saturating flashes of about 16

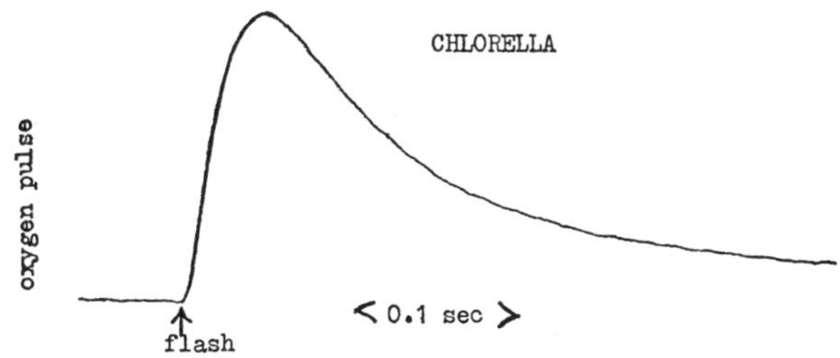

FIGURE 1 Recording of the current of an oxygen polarograph, caused by illumination with a 15 μsec xenon flash of a thin layer of *Chlorella* deposited on a horizontal platinum electrode. The current is proportional to the rate of oxygen diffusion to the electrode.

μsec half-width. A recording of the electrode current after a flash is given in Fig. 1. The current increases to 70% of the maximum in 0.02 sec, and decreases slowly. The shape of the curve is not markedly influenced by the recording system, which has a frequency response of several hundred hertz. For low intensity flashes the value at the maximum of the curve was proportional to the light intensity, and the curves had "similar shapes"; that is, the curves could be transformed into each other by multiplying by a constant factor. This indicates that the value of the maximum can be taken as proportional to the amount of oxygen produced by the flash. The current starts increasing within a millisecond or less after the flash and the 0.02 sec rise time may be caused largely by diffusion of oxygen through the thin algal layer. The relatively slow decrease of a half-time of 0.1 sec is presumably caused by back diffusion of oxygen from a distance. Only one turnover of Q is to be expected by a 16 μsec flash, since the reoxidation of Q^- requires about 200 μsec (Zankel and Kok, 1970). Thus the value of the maximum of the oxygen pulse in a saturating flash may be assumed to be proportional to the number of active reaction centers, if the Z is in the steady state, that is, the concentrations of the Z^{n+}'s ($n = 0$, 1, 2, 3) are equal (see below), and if Q is in the oxidized state.

RESULTS AND DISCUSSION

In Fig. 2 is shown an experiment which was performed by L. Vorst. A number of saturating flashes is given, then a period of darkness of about 29 sec, then again a number of flashes. The algal species used was *Chlorella vulgaris*. The oxygen pulses after the period of darkness show a fourfold periodicity; the first, fifth, etc., pulses are smaller than the others. This periodicity has been studied by Joliot et al. (1969) and Kok et al. (1970), and was explained by the latter authors by the hypothesis that at each reaction center four oxidizing equivalents, each produced by one quantum, are needed in order to produce one oxygen molecule. In the redox states Z^{n+} ($n = 0, 1, 2, 3$), each quantum exciting C causes the reaction $Z^{n+} \rightarrow Z^{(n+1)+}$. The concentration of Z^{4+} is, in the dark, negligible because Z^{4+} reacts rapidly with water: $Z^{4+} + 2H_2O \rightarrow Z + O_2 + 4H^+$. The concentrations of the Z^{n+}'s after a dark period are reflected in the periodic responses. After some time the variations in the oxygen pulses damp out and a "steady state" results in which the Z^{n+}'s have equal concentrations.

After a new sequence of flashes a dark period of equal length is given but at the beginning of a second dark period of 29 sec the inhibitor of photoreaction II,

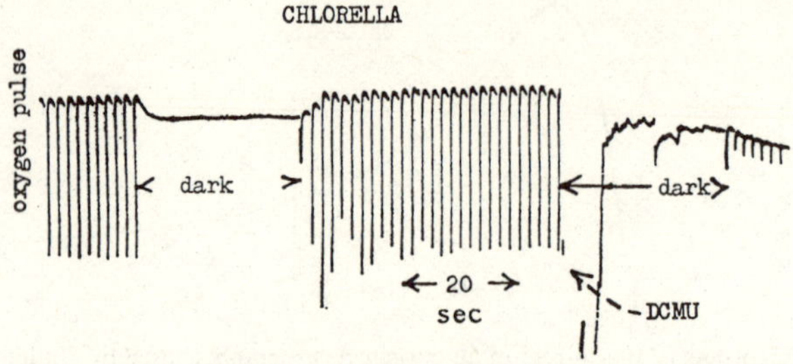

FIGURE 2 Recording of current produced by oxygen pulses in a series of xenon flashes spaced by two dark periods of 29 sec. At the beginning of the second dark period DCMU is added to a concentration of 10 μM. After DCMU addition only the first pulse yields oxygen (see text). In this recording only the AC component of the current is recorded in order to reduce disturbances caused by the addition of DCMU.

DCMU, is added to a concentration of 5×10^{-5} M. The deflections in this dark period are disturbances on the electrode by the addition of DCMU. Fig. 2 shows that only the first flash causes oxygen evolution. Oxygen evolution in the other flashes is inhibited. The remaining deflections are disturbances of the flash. The experiment was repeated many times also with shorter dark periods. Always only the first pulse was not reduced or only slightly reduced; the other oxygen pulses were partly or completely inhibited. If partly inhibited, the ratios of the pulses after the first one were the same as those after the first pulse after a dark period without inhibitor. The simplest and most plausible interpretation of this experiment is that DCMU prevents the reoxidation of Q by the oxidant A, presumably plastoquinone (Amesz, 1964), and thus only the first excitation of C is active.

Within 10 sec after the flash, whether in the presence or absence of DCMU, the fluorescence yield of chlorophyl a_2 and thus the concentration of Q^- has dropped to a rather low value (Bennoun, 1970), so that even in the presence of DCMU we may expect that oxygen pulses will be caused by flashes spaced at 10 sec. In the absence of DCMU with flashes spaced at 10 sec appreciable oxygen pulses occur, but we found that in the presence of 5×10^{-5} M DCMU the oxygen pulses in the steady state virtually were reduced to zero. This can be explained by the hypothesis that Q^-, in the absence of DCMU, reacts most rapidly with A, but in its presence cannot react with A and reacts back with Z^{3+} or another of the Z^{n+}'s, thus preventing accumulation of Z^{3+}. This interpretation is in accordance with the observation (Bertsch et al., 1963; Clayton, 1969; Bennoun, 1970) that in the presence of DCMU, luminescence, presumably resulting from the back reaction of Z^{n+} with Q^- (Bennoun, 1970), still occurs.

A long known phenomenon is the slow decrease in chlorophyll a_2 fluorescence yield in algae, after the rapid increase in fluorescence yield due to reduction of Q upon onset of illumination. This decrease takes place in a time of a few minutes and was shown to occur only in light II. It was interpreted as a decrease in activity of system II (Duysens and Sweers, 1963). The system II activity could be restored in about 10 sec by light I (or in some species of algae by a longer period of darkness). This restoration of activity also occurred in the presence of DCMU by excitation of system I (Duysens and Talens, 1969). Thus the changes between pigment states are not produced by electron transfer between Q^- and system I. Pigment state I appeared to be produced by the accumulation of one or more components between Q and system I in the oxidized state, pigment state II by accumulation in the reduced state. The inactivation of system II was originally attributed to an inactivation of Q by a transition of the form Q into a photoinactive form Q' which quenched the fluorescence (Duysens and Sweers, 1963; Duysens and Talens, 1969). Bonaventura and Myers (1969) concluded from parallel measurements of oxygen rate and fluorescence yields that in prolonged light II the photosynthetic activity of system II decreased (in *Chlorella* by 9%) and that of system I increased. They proposed the hypothesis that changes in activities were caused by a change in distribution of the absorbed energy over the two systems, rather than by a decrease in activity of reaction center II.

The following experiments, reported earlier (Duysens, 1969, 1970), show that indeed the reaction center II is not inactivated in light II. These experiments were performed by A. S. Kwak and Th. E. van der Schatte Olivier. *Chlorella pyrenoidosa* was used. Within the precision of measurement of about 1%, the oxygen pulse produced by a saturating xenon flash of 16 μsec half-width was found to be the same in the two states which were caused by prolonged illumination with light I and II, respectively. Bonaventura and Myers have called these states light I and light II states, but since the shift to the light I state also occurs in darkness (Duysens and Sweers, 1963) and, in the presence of DCMU also is caused by light II (Duysens and Talens, 1969), the terms pigment states I and II are preferable (Duysens, 1969, 1970).

To avoid the "Joliot periodicity" the experiments were done under the steady state of the Z's. In state I a background of light I was used, and in state II, immediately after switching off the light II needed for producing state II, a short period of illumination with light I was given, sufficient to secure oxidation of all Q^-, but too short to cause a back transition to state I. In half-saturating flashes the oxygen pulses were 7–10% smaller in pigment state II than in pigment state I, showing that fewer quanta became available in the reaction centers in state II. Extending Bonaventura and Myers' interpretation, the hypothesis was proposed (Duysens, 1969, 1970) that the pigment state I to II transition is caused by a change in the thylakoid membrane by which the pigments of the two reaction centers are moved closer towards each other. This causes a decrease in the fluorescence yield (mainly from system II) and an increase in system I trapping. This hypothesis is supported by the observation that in *Chlorella* the fluorescence yield increase in the xenon flash adjusted to a just saturating intensity for oxygen pulses is about 10% higher in pigment state I than in pigment state II.

Independently analogous conclusions were drawn by Murata (1969) concerning decreases in chlorophyll a_2 fluorescence in spinach chloroplasts. In Murata's chloroplast preparation the decrease in maximum fluorescence yield could not be brought about by illumination with light II, but rather by decreasing the concentration of magnesium or other ions, or by producing a so-called "high energy state" (see also Wraight and Crofts, 1970). These two changes in chloroplasts presumably have different causes. In vivo a high energy state may be brought about by illumination with light I, but in vivo the effect of illumination with light I is a higher maximum fluorescence yield (pigment state I) and not a smaller one as in chloroplasts.

It is thus quite possible that the shifts of the pigment systems occurring in intact cells have a different cause from those observed in chloroplasts, although conformational changes and changes of charged ions on the membrane (Vredenberg, 1970) or of the membrane potential may play a role in both chloroplasts and intact cells. (See recent reviews by Myers, 1971; Govindjee and Papageorgiou, 1971.)

This investigation was supported in part by the Netherlands Foundation for Biophysics, financed by the Netherlands Organization for the Advancement of Pure Research (ZWO).

REFERENCES

AMESZ, J. 1964. *Biochim. Biophys. Acta.* **79**:257.

BENNOUN, P. 1970. *Biochim. Biophys. Acta.* **216**:357.

BERTSCH, W. F., J. B. DAVIDSON, and J. R. AZZI. 1963. *Natl. Acad. Sci. Natl. Res. Counc. Publ.* **1145**: 701.

BONAVENTURA, C., and J. MYERS. 1969. *Biochim. Biophys. Acta.* **189**:366.

CLAYTON, R. K. 1969. *Biophys. J.* **9**:60.

DUYSENS, L. N. M. 1964. *Prog. Biophys. Mol. Biol.* **14**:1.

DUYSENS, L. N. M. 1969. Structure, Function and Control Mechanism in Photosynthetic Organelles. Gordon Research Conference. Plymouth, N.H.

DUYSENS, L. N. M. 1970. International Conference on the Photosynthetic Unit, Gatlinburg, Tenn. (Abstr. A4).

DUYSENS, L. N. M., and H. E. SWEERS. 1963. *In* Studies on Microalgae and Photosynthetic Bacteria. Special issue of *Plant Cell Physiol.* Japanese Society of Plant Physiologists, editors. University of Tokyo Press, Tokyo. 353.

DUYSENS, L. N. M., and A. TALENS. 1969. *In* Progress in Photosynthesis Research. H. Metzner, editor. Laupp'sche Buchhandlung, Tübingen, W. Germany. **2**:1073.

GOVINDJEE and G. PAPAGEORGIOU. 1971. *Photophysiology.* **6**:1.

JOLIOT, P., G. BARBIERI, and R. CHABAUD. 1969. *Photochem. Photobiol.* **10**:309.

KNOX, R. S. 1968. *J. Theor. Biol.* **21**:244.

KOK, B., B. FORBUSH, and M. MCGLOIN. 1970. *Photochem. Photobiol.* **11**:457.

MURATA, N. 1969. *Biochim. Biophys. Acta.* **189**:171.

MYERS, J. 1971. *Annu. Rev. Plant Physiol.* **22**:289.

RABINOWITCH, E. I. 1951. Photosynthesis and Related Processes. Interscience Publishers Inc., New York. **2**(Pt. 1):823.

VREDENBERG, W. J. 1970. *Biochim. Biophys. Acta.* **223**:230.

WRAIGHT, C. A., and A. R. CROFTS. 1970. *Eur. J. Biochem.* **17**:319.

ZANKEL, K. L., and B. KOK. 1970. International Conference on the Photosynthetic Unit, Gatlinburg, Tenn. (Abstr. B3).

Chloroplast Structure

STRUCTURAL DIFFERENTIATION OF STACKED AND UNSTACKED CHLOROPLAST MEMBRANES

Freeze-Etch Electron Microscopy of Wild-Type and Mutant Strains of *Chlamydomonas*

URSULA W. GOODENOUGH and L. ANDREW STAEHELIN

ABSTRACT

Wild-type chloroplast membranes from *Chlamydomonas reinhardi* exhibit four faces in freeze-etchreplicas: the complementary Bs and Cs faces are found where the membranes are stacked together; the complementary Bu and Cu faces are found in unstacked membranes. The Bs face carries a dense population of regularly spaced particles containing the large, 160 ± 10 A particles that appear to be unique to chloroplast membranes. Under certain growth conditions, membrane stacking does not occur in the *ac*-5 strain. When isolated, these membranes remain unstacked, exhibit only Bu and Cu faces, and retain the ability to carry out normal photosynthesis. Membrane stacking is also absent in the *ac*-31 strain, and, when isolated in a low-salt medium, these membranes remain unstacked and exhibit only Bu and Cu faces. When isolated in a high-salt medium, however, they stack normally, and Bs and Cs faces are produced by this in vitro stacking process. We conclude that certain particle distributions in the chloroplast membrane are created as a consequence of the stacking process, and that the ability of membranes to stack can be modified both by gene mutation and by the ionic environment in which the membranes are found.

INTRODUCTION

The structural organization of chloroplast membranes in green algae and in higher plants is distinct from that of all other membranes. In fixed and sectioned material an individual chloroplast membrane does not appear unusual, being approximately 70 A wide and exhibiting a trilaminar structure. However, in preparations of whole chloroplast membranes that are either shadowed or negatively stained, unique arrays of particles are seen (3, 28, 40, 42). Moreover, when chloroplast membranes are freeze-cleaved or freeze-etched, faces are revealed that exhibit not only a higher particle density than other membranes (8), but also a distinctive type of large particle found in no other membrane that has yet been studied. These large particles have been termed "quantasomes" or, more recently, "quantasome cores" by Park and coworkers (9, 42, 43) who have suggested that the particles might be the morphological counterpart of some functional unit of photosynthesis (41, 44).

Chloroplast membranes also differ from most membranes in that they possess the capacity to fuse together over long distances, forming "grana" in higher plants and smaller "stacks" in green algae. Membrane fusion occurs in myelin (47) and in certain specialized regions of the plasma membrane such as tight junctions (16, 20, 49); however, most membranes do not make contact over extensive distances under physiological conditions.

The two distinctive structural properties of chloroplast membranes—their particle population and their ability to fuse—are the subject of the present investigation. We have studied the appearance of chloroplast membranes from three different strains of the unicellular green alga, *Chlamydomonas reinhardi*, by electron microscopy of thin-sectioned and freeze-etched material. Freeze-etched replicas from wild-type membranes exhibit the same types of faces and particles as do replicas of membranes from higher plant chloroplasts. However, our interpretation of such replicas differs in certain important respects from those that have been offered by other investigators. We find that stacked membranes possess a freeze-etch morphology that is distinct from that of unstacked membranes. Parallel studies of two mutant strains, *ac-5* and *ac-31*, strongly support this conclusion, for under certain growth or experimental conditions, these strains possess exclusively unstacked membranes and exhibit only the morphology characteristic of unstacked membranes in freeze-etched preparations.

A comparison of the membrane structure of these strains has enabled us to draw the following conclusions: (*a*) the characteristic arrays of large particles found within chloroplast membranes are produced when stacking occurs and are absent from unstacked membranes, whether or not these membranes possess the potential for stacking; (*b*) a dramatic reduction in the number of large particles is not accompanied by any loss of photosynthetic capacity as measured by CO_2 fixation or photosynthetic electron-transport reactions at high light intensities; (*c*) chloroplast membrane stacking, and the concomitant formation of distinct membrane faces, is mediated by ionic interactions between membranes; (*d*) the chloroplast membranes appear to be differentiated into regions that can and cannot stack.

MATERIALS AND METHODS

Culture of the Organisms

Wild-type (strain 137c) and mutant (strains *ac-5* and *ac-31*) cells were grown in 3-l volumes of minimal medium (50) or minimal medium supplemented with 3% sodium acetate contained in 4-l Erlenmeyer flasks and agitated with magnetic stirrers. The light intensity from daylight fluorescent lamps was 2000 lux, and the temperature was maintained at 26°C. Cultures were harvested in the logarithmic phase of growth.

Preparation of Chloroplast Fragments

A sample of intact cells was first fixed for thin-section electron microscopy by methods previously described (23, 31). The remaining cells were washed and broken gently in a French pressure cell (35–50 kg/cm^2). The broken preparation was subjected to differential centrifugation as described by Surzycki (51) to eliminate whole cells, starch, and small membrane fragments, but no attempt was made to isolate whole chloroplasts; the final "chloroplast fraction" therefore contained some protoplasts and chloroplasts and a predominance of large chloroplast membrane fragments.

All procedures were carried out at 4°C. In most cases the buffers used were those described by Surzycki (51): for washing, 0.05 M Tris HCl (pH 8), 0.25 M sucrose, 0.001 M Mg EDTA, and 0.02 M $MgSO_4$; for breaking and subsequent centrifugations, 0.05 M Tris HCl (pH 8), 0.5 M sucrose, 0.02 M $MgSO_4$, and 0.25% bovine serum albumin. This second solution will be referred to in the text as high-salt buffer.

To isolate unstacked *ac-31* chloroplast membranes, cells were washed, broken, and centrifuged in 0.05 M Tricine NaOH, pH 7.3 (19); this will be referred to as low-salt Tricine buffer. To prepare unstacked wild-type membranes, cells were washed, broken, and centrifuged as usual in high-salt buffer, and the fragments were then subjected to five washings in the low-salt Tricine buffer over a 1 hr period.

The final chloroplast membrane fraction was suspended in 11 ml of cold high- or low-salt buffer, and 4 ml of glycerol were slowly added over a 30 min period. After an additional 30 min at the final glycerol concentration (ca. 27%), the membranes were spun down at 5000 *g* for 3 min. A portion of the pellet was used for preparing freeze-etch specimens; these were rapidly frozen in Freon 12 held at −150°C. The remainder of the pellet was used for thin-section microscopy. The fragments were usually washed once in 0.15 M potassium phosphate buffer, pH 7.4, containing 0.5 M sucrose and 27% glycerol. This step removes Tris, which is an unsuitable vehicle for glutaraldehyde fixation (13). The fragments were then resuspended in this solution, and an equal volume of 4% glutaraldehyde in the same buffer was added. After a 2–3 hr fixation, they were washed in the buffer without glycerol and postfixed in 1% OsO_4 in the same buffer for 45 min. They were then washed in the phosphate buffer without sucrose or glycerol, dehydrated rapidly in ethanol, and introduced into Epon-Araldite (31) without intervening propylene oxide. They were transferred to fresh monomer after 2 days, subjected to evacuation at 80°C, and embedded at 80°C for 3 days.

To fix the low-salt preparation of *ac-31* fragments, the final pellet was suspended in the Tricine buffer containing 27% glycerol to which an equal volume of 0.004 M potassium phosphate buffer, pH 7, containing 4% glutaraldehyde was added. The remaining procedures were performed in the dilute phosphate buffer. The unstacked wild-type fragments were suspended in the Tricine buffer, and a commercial 20% stock solution of Taab (Reading, England) glutaraldehyde was added to a final concentration of 2%. The remaining procedures were performed in the dilute phosphate buffer. Both procedures for fixing in the presence of Tricine give unsatisfactory thin-section images as discussed in the text.

The freeze-etch replicas were prepared at −100°C according to the method described by Moor and Mühlethaler (38). However, to avoid postcleaving changes of the exposed membrane surfaces, the specimens were replicated within 3–5 sec of the last pass of the microtome knife.

Thin sections were examined with a Hitachi HU-11C and replicas were examined with a Philips 300 electron microscope.

Measurement of Photosynthetic Parameters

Total chlorophyll and chlorophyll a:b ratios were estimated by a modification (2) of the method of Mackinney (37). Cells were counted in a hemacytometer. Carbon dioxide fixation was measured as the amount of $^{14}CO_2$ fixed by whole cells, by using a method described elsewhere (21). The activity of PS II[1] was measured as the light-induced reduction of DPIP from water, and PS I activity was measured as the light-induced reduction of NADP from reduced DPIP. The total operation of the photosynthetic electron-transport chain was measured as the light-induced reduction of NADP from water. Assays were made using chloroplast fragments prepared by the sonic disruption of cells. References to methods will be found in reference 21.

Measurement of Particle Sizes

Particle sizes were determined from micrographs enlarged to 160,000 × and viewed through an 8 × objective lens equipped with a micrometer grating. The width of the shadow of a given particle was measured over the shadowed half of the particle. Where the shadow boundary appeared fuzzy or irregular, a minimum width value was always taken. Between 200 and 400 particles were measured in large continuous areas on at least two different micrographs for each histogram of particle size distribution (Fig. 9). Histograms of "B"-type faces are more easily obtained and are therefore used to define changes in particle populations of freeze-etched membranes.

RESULTS

Wild-Type Chloroplast Membranes

The organization of chloroplast membranes in whole cells of wild-type *C. reinhardi* is shown in Fig. 1. Stacks of 2–10 thylakoids, or discs, lie within the chloroplast stroma, as described in detail else-

---

[1] Abbreviations used: PS II, photochemical system II; PS I, photochemical system I; DPIP, 2,6-dichlorophenylindophenol; NADP, nicotinamide adenine dinucleotide phosphate.

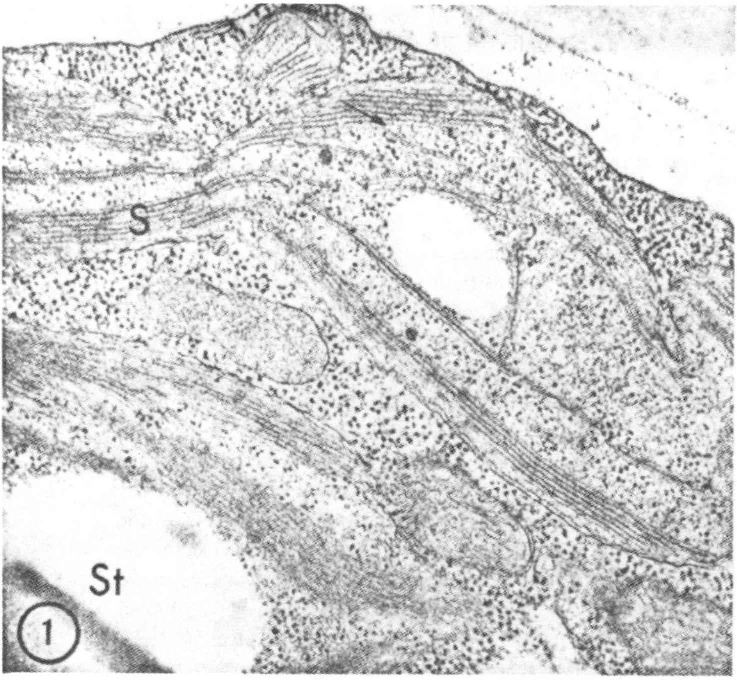

Figure 1 Portion of a wild-type *C. reinhardi* cell grown mixotrophically, showing two lobes of the chloroplast. Thylakoids are usually found in stacks *(S)* of 2-7 thylakoids. *Arrow* indicates region where single thylakoid joins a larger stack. Starch is shown at *ST.* X 48,000.

where (22). The apposed membranes in a stack apparently make direct contact, for the width of two stacked membranes is consistently twice that of a single, unstacked membrane in sectioned material (22).

C. reinhardi is impermeable to glycerol, and attempts to produce freeze-etch replicas from unglycerinated cells were unsatisfactory. We therefore isolated large chloroplast fragments and suspended them in a buffered 27% glycerol solution. Thin-section microscopy of such fragments (Fig. 2) reveals that the chloroplast thylakoids retain their stacked configuration even after disruption, extensive washing, and chemical fixation, as has also been observed for preparations of higher plant chloroplast fragments (32, 56). In other words, the association between chloroplast membranes appears to be a very strong and durable one.

Fig. 3 shows at low magnification, a typical freeze-etched replica obtained from unfixed wild-type chloroplast membranes, and Figs. 4–8 show the characteristic membrane faces of such replicas at higher magnifications. Experimental results of others (43, 45, 53, 55) have supported the theory (6, 9) that the fracture plane travels preferentially along the interior of the membrane in such a way that two complementary faces are exposed. As this theory is consistent with our observations, it forms the basis of our interpretations of freeze-etch replicas. We have called the two complementary faces of chloroplast membranes B and C, since they are equivalent to the B and C faces of Branton and Park (9). However, we find evidence for two different types of B and C faces, namely, those that occur within areas of membrane that are stacked and those that occur within areas of membrane that are unstacked. We shall therefore refer to the faces that derive from stacked membranes as Bs and Cs, and those that occur within unstacked membranes as Bu and Cu. These faces and the particles associated with them are described below.

The Bs Face

The Bs is the most distinctive face of chloroplast membranes and corresponds to the B face of Branton and Park (see Figs. 5, 6, and 9 of reference 9). It is characterized by a dense population of fairly regularly spaced particles. We have measured the size of more than 400 of these particles, and the results are plotted in Fig. 9 *a*. It is evident from this histogram, as well as from a careful inspection of Figs. 3 and 5, that the particles are not uniform in size. The majority of the particles seem to fall into three major size categories: 105 ± 10 A, 130 ± 10 A, and 160 ± 10 A. It is most likely that our 160 ± 10 A particle corresponds to the 175 A "quantasome" of Branton and Park (9). It often appears to be composed of subunits, although we should point out that the smaller particles on the Bs face also occasionally appear to be made up of subunits. Particles in the 160 ± 10 A category account for some 22% of the particles on the wild-type Bs membrane face (Fig. 9 *a*). We cannot exclude the possibility that at least some of the smaller particles found on this face might arise from larger particles that are torn apart during the cleaving process, as discussed more fully in a later section.

We have chosen to characterise our membrane faces by particle-size distribution data rather than by estimates of particle density. Use of the latter parameter requires making a choice of which membrane area to count; as is evident in the figures in this paper, particle density varies considerably from one region of a given face to the next. Very marked differences in particle density, such as the difference between Bs faces and the Bu faces described below, are sufficiently obvious to render quantitation unnecessary.

The Bs face usually appears to have a very smooth background matrix. At low shadowing angles, however, the matrix between the particles reveals large numbers of small, shallow depressions, most of which have diameters of 20–70 A (Fig. 5, *arrows*).

As indicated in Fig. 3, Bs surfaces are found exclusively in areas of the replica that derive from regions of membrane stacking.

The Cs Face

The Cs membrane face (Figs. 3, 4, and 6) is—like the Bs face—only seen in regions where the thylakoid membranes are in a stacked configuration (Fig. 3). The Cs face appears to be quite uneven, largely because it contains a considerable number of large holes, 50–160 A in diameter (Fig. 6, *arrows*), and, to a lesser extent, because it carries a moderate number of small particles. Most of the particles range in size from 30 to 100 A (Fig. 9 *c*). Our Cs membrane faces appear to correspond to the faces labeled *A* on the micrographs of Branton and Park (9) and Arntzen et al. (3), but they do not correspond to the surfaces labeled *A* in the diagrams of these authors (see Discussion).

The Bu Face

Bu faces are typical of thylakoid membrane regions that do not take part in the stacking process (Figs. 3, 7). They are continuous with the Bs faces (Fig. 3) and appear to share the same background matrix. In the Bu areas, however, the matrix is marked by numerous depressions (Fig. 7, *arrows*)

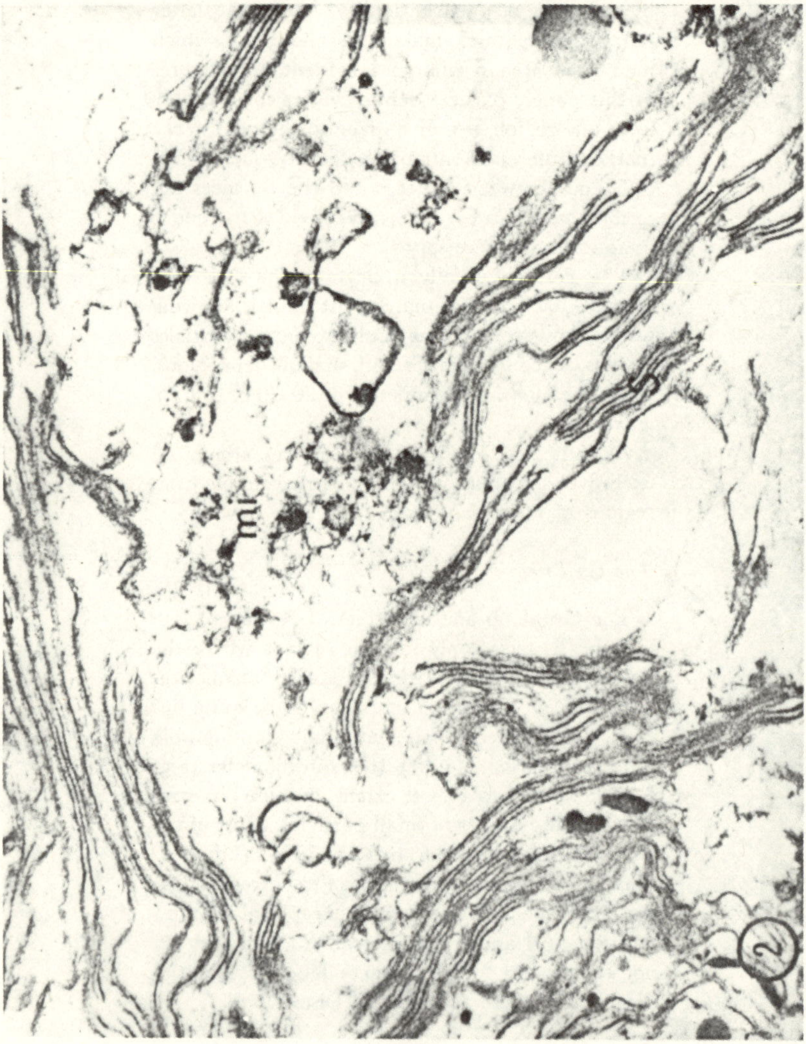
mi
2

FIGURE 2 Isolated chloroplast membranes from wild-type cells, glutaraldehyde-fixed in 27% glycerol in high-salt buffer. Thylakoids retain their stacked configuration (*S*), even though the chloroplast envelope has ruptured and stroma contents have been lost. Microsomes (*mi*) and other membranes are also present in the preparation. × 48,000.

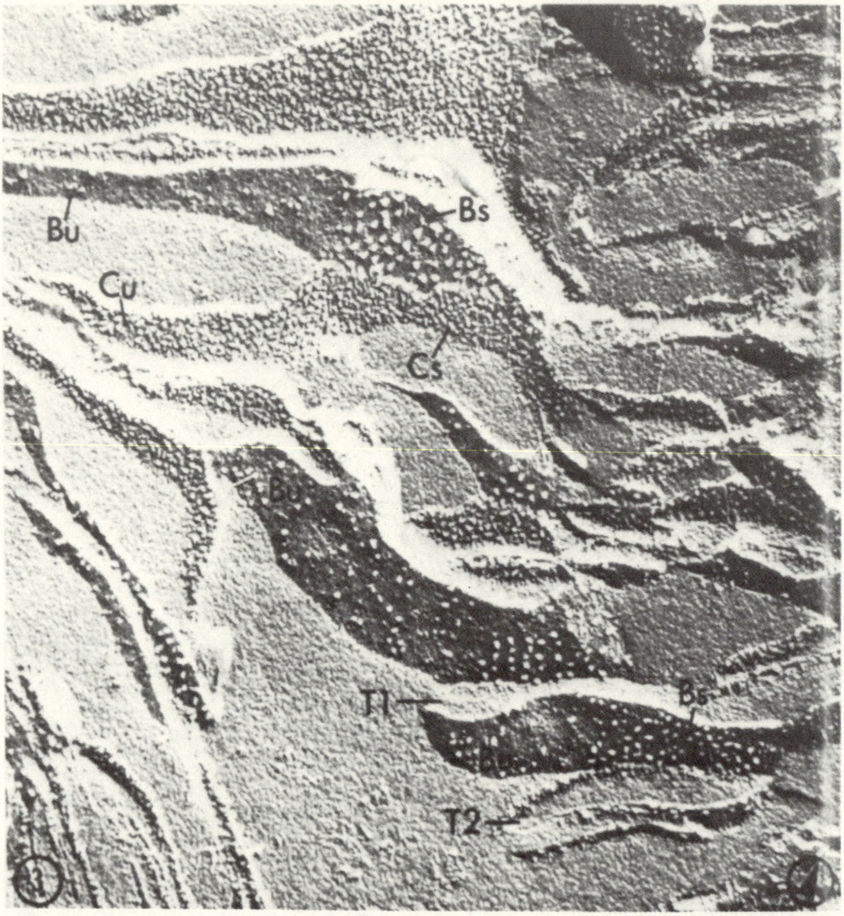
Bu
Bs
Cu
Cs
Bu
T1
Bs
Bu
T2

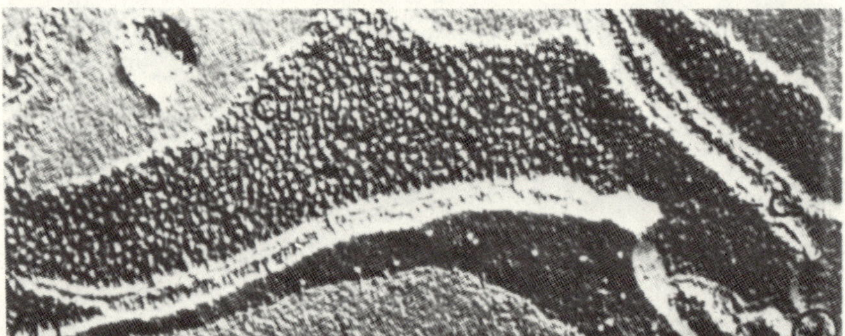
Cu

FIGURE 3 Freeze-etched isolated chloroplast membranes from wild-type cells, suspended in 27% glycerol in high-salt buffer. Four types of split inner membrane faces can be distinguished; they have been labeled *Bs*, *Bu*, *Cs*, and *Cu*, for reasons discussed in the text. Faces Bs and Bu are continuous with each other and face outwards with respect to the lumen of the thylakoids; Cs and Cu are also continuous but face inwards (See Figs. 4 and 21). *T1* and *T2* are two adjacent thylakoids. Where they come together to form a stacked region, a typical Bs (B stacked) face is seen. The adjoining Bu (B unstacked) face, which leads up to the edge of the stacked region, is characteristic of unstacked membrane areas. The relationship between Cs and Bs faces in stacked membrane regions is apparent in the upper half of the picture. The Cs face changes into a Cu face where the membranes become unstacked. Faces Bs, Cs, Bu, and Cu are shown in greater detail in Figs. 5–8. Circled triangle in lower right shows the angle of shadowing. × 80,000.

FIGURE 4 Micrograph demonstrating both the continuity between Cs and Cu faces and their fine-structural differences. The *Cs* face appears as a rough, irregular surface with predominantly small particles (See Fig. 9 *c*) distributed between large depressions. The *Cu* face has a more regular appearance with numerous, medium-sized particles (Fig. 9 *d*) protruding from a relatively smooth background in which few holes can be detected. Wild-type chloroplast preparation suspended in 27% glycerol in high-salt buffer. Circled triangle in lower right shows the angle of shadowing. × 100,000.

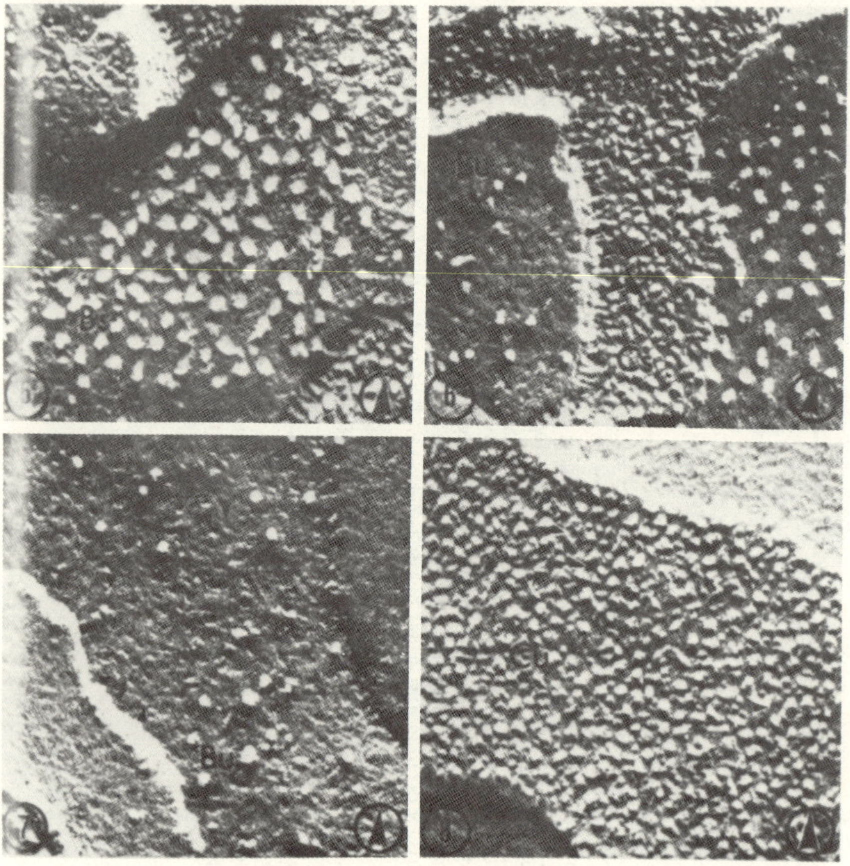
Bs
Bu
Bu
Cu

FIGURES 5–8 The four membrane faces *Bs*, *Cs*, *Bu*, and *Cu* from wild-type chloroplast preparations suspended in 27% glycerol in high-salt buffer. The suggested relationships of these faces to one another are indicated schematically in Fig. 23 *B*, and are discussed in detail in the text. Fig. 5, A *Bs* face from a replica area with a low shadowing angle, demonstrating the fine-structural details in the background matrix. Numerous small holes (*arrows*), with diameters of 20–80 A, interrupt the smooth matrix. A population of regularly spaced large particles (Fig. 9 *a*) is also seen; some of these exhibit a substructure. Fig. 6, A *Cs* face, exhibiting numerous large holes (*arrows*) that are distributed between the particles (Fig. 9 *c*) of the background material. On the right edge of the Cs face it is seen that the tips of the large particles of the underlying *Bs* face are approximately level with the tips of the particles on the Cs face. Some of the large Bs particles at the edge seem to be continuous with and even identical with, a part of the background material making up the Cs face. Fig. 7, A *Bu* face, which carries few heterogenously sized particles (Fig. 9 *b*) and exhibits many large depressions (*arrows*) within its relatively smooth background matrix. Fig. 8, A *Cu* face, covered with numerous, medium-sized particles (Fig. 9 *d*); these are set in a smooth background material in which only a few holes (*arrows*) can be recognized. Circled triangle in lower right shows the angle of shadowing. Figs. 5–8 × 160,000.

that resemble pockmarks and appear much larger (50–120 A in diameter) and deeper than the depressions in the Bs matrix (Fig. 5, *arrows*). The Bu face also carries a relatively sparse population of particles. The average particle size is smaller than that of a Bs face (compare Fig. 9 *b* with Fig. 9 *a*). This difference is mainly due to a sharp decrease in the relative number of 160 $\pm$ 10 A particles; to a lesser extent it is also due to a decrease in the number of 130 $\pm$ 10 A particles and a proportional increase in the number of particles with diameters of 80–115 A. Notice also the shift in the average size of the smaller particles from 105 to 95 A as one moves from a Bs to a Bu face. Branton and Park (9) refer to the Bu face as a B face from which the "quantasome" particles have been torn away, leaving large holes. In Fig. 5 of reference 9, a large Bu face is present but it is not specifically distinguished from the adjacent "B" face.

The Cu Face

Cu faces are also characteristic of unstacked thylakoid membranes (Fig. 3). They are continuous with Cs membrane faces (Figs. 3, 4) in the same way that Bu and Bs faces are continuous with each other. The background matrix of the Cu face is smoother than that of the Cs face and carries only infrequent depressions (Fig. 8, *arrows*); for this reason, one can more readily visualize its population of densely packed particles (Figs. 4 and 6). The diameters of these particles are difficult to determine because of their close packing and frequently irregular form, and thus we have only attempted to determine the particle size distribution of wild-type Cu faces. The resulting histogram (Fig. 9 *d*) indicates that the particle sizes on a Cu face do not differ significantly from those seen on a Bu face (Fig. 9 *b*), although more small particles ($<$80 A in diameter) are found on a Cu face and the peaks at 95, 125, and 155 A are less pronounced. Our Cu faces correspond to the C faces of Branton and Park (9) and of Arntzen et al. (3).

To confirm our identification of stacked and unstacked membrane faces, and to obtain information on the possible nature of the 160 $\pm$ 10 A particles, we have extended our study to two strains of *C. reinhardi* that carry mutations affecting the stacking properties of their chloroplast membranes.

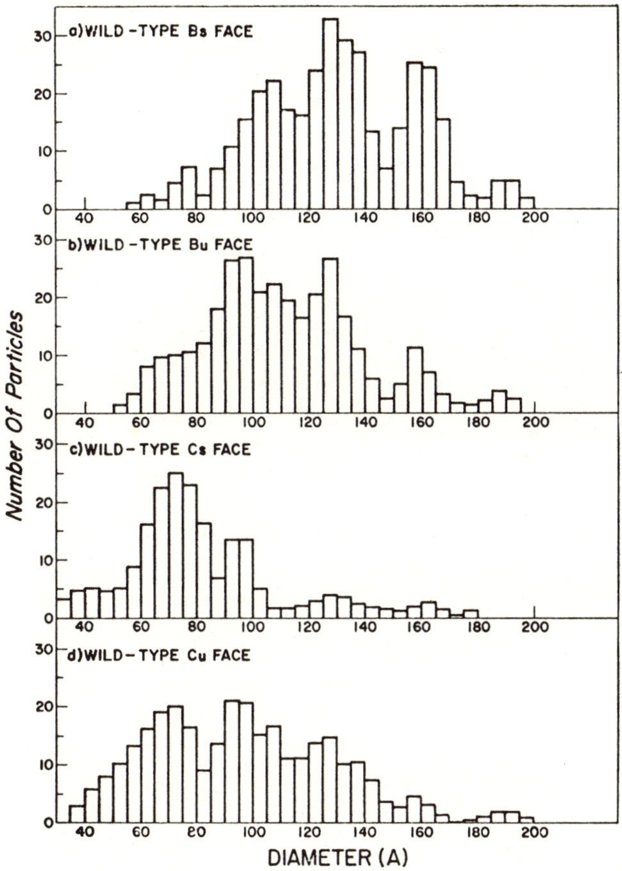

Figure 9 Histograms of particle sizes found on various membrane faces in freeze-etch preparations of *C. reinhardi* chloroplast membranes. The histograms are labeled *a-k*, and each is identified by individual labels.

The Chloroplast Membranes of ac-5

The *ac-5* mutation in *C. reinhardi* produces cells that are chlorophyll deficient, and particularly chlorophyll *b* deficient, in comparison to wild-type cells (Table I). The mutant cells can be grown either phototrophically on a minimal medium or mixotrophically in the light on a minimal medium supplemented with acetate; under both growth conditions the cells are equally pigment deficient or, if anything, the phototrophic cells tend to be more pigment deficient than the mixotrophic cells (Table I). The photosynthetic capacity of *ac-5* cells

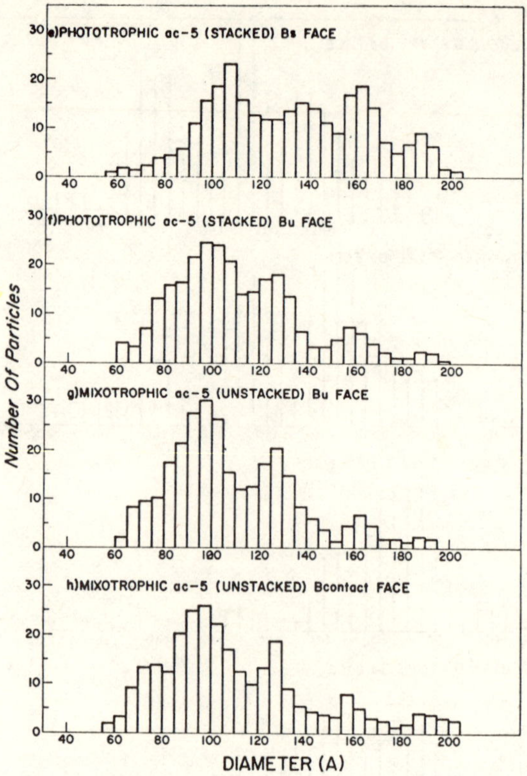

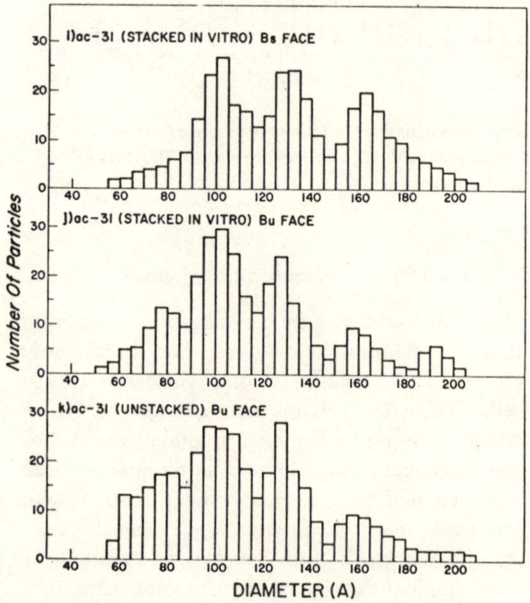

For legend see page 163.

is also comparable under both growth conditions, whether measured as the ability of whole cells to fix CO_2 in the light or as the ability of isolated chloroplast fragments to photoreduce DPIP or NADP (Table I). Thus, there is no apparent difference in PS I or PS II activity under the two growth conditions. It should be noted that these assays were all performed at light intensities that are saturating for wild-type cells; we do not yet have information on the photosynthesis of *ac-5* at low light intensities, although such experiments are now being carried out by R. P. Levine.

When the fine structure of *ac-5* cells is examined, a striking difference is found between phototrophically grown and mixotrophically grown cells. Figs. 10 and 11 show, respectively, a phototrophic cell and chloroplast fragments from a phototrophic cell. It is evident that most of the chloroplast stacks contain only two thylakoids and that there are many more single thylakoids present than are found in wild-type chloroplasts (compare with Figs. 1 and 2). Nonetheless, membrane stacking unquestionably occurs. In contrast, mixotrophic *ac-5* cells are incapable of any chloroplast membrane stacking (Fig. 12). The membranes traverse the chloroplast stroma in an apparently independent manner, and no examples of fusion have ever been found. When chloroplast fragments are isolated from such cells, moreover, the membranes remain unstacked (Fig. 13), although, very occasionally, one encounters short regions where the membranes give the appearance of having fused (Fig. 13, *arrows*).

Freeze-etch replicas of isolated phototrophic and mixotrophic *ac-5* chloroplast membranes are shown in Figs. 14–16. Phototrophic *ac-5* replicas exhibit the four types of membrane faces, Bs, Cs, Bu, and Cu, found in wild-type replicas (Fig. 14). The histograms of particle sizes on the B faces also correspond to those of the wild type (compare Figs. 9 *e* and *f* with Figs. 9 *a* and *b*). In contrast, only Bu and Cu faces are found in replicas from the unstacked, mixotrophic *ac-5* cells (Figs. 15–16). The distribution of particle sizes on the Bu faces of mixotrophic and phototrophic *ac-5* membranes is comparable (compare Fig. 9 *g* with Fig. 9 *f*). The Bu faces of mixotrophic *ac-5* cells tend to carry a somewhat higher density of particles than the Bu faces of wild-type cells (compare Figs. 3 and 16), but it is difficult to express this observation quantitatively since, as we have pointed out earlier, particle densities vary so markedly from one membrane area to another.

Occasionally, small regions are encountered in replicas of mixotrophic *ac-5* cells that, we believe, correspond to the short regions of membrane contact occasionally seen in thin section (Fig. 13). One such region is included in Fig. 15 (*arrow*): two membranes are seen to come together, and in the region of contiguity the density of particles increases on the exposed membrane face. When examined by eye, such regions appear to lack the arrays of large particles typical of truly stacked membranes, and this impression is confirmed when particle size measurements are made. As seen in Fig. 9 *h*, the distribution of particle sizes in the regions of membrane contact remains the same as on Bu faces. No Bs particle distribution is created and, in particular, no increase is observed in the relative number of 160 ± 10 A particles. Thus freeze-etching allows a clear distinction between true chloroplast membrane stacking and regions of close membrane association.

The Chloroplast Membranes of ac-31

A detailed study of the *ac-31* strain has been published elsewhere (21). Table I summarizes the photosynthetic properties and pigment content of this strain, all of which are seen to be comparable to those of the *ac-5* strain. It should be noted that, since both strains are pigment deficient, both give rates of photosynthetic reactions that are high compared with those of the wild type when calculated on a chlorophyll basis, and low compared with those of the wild type when calculated on a cell basis. It is thus difficult to make meaningful comparisons between the photosynthetic capacities of these cells and wild-type cells, as discussed more fully elsewhere (21).

The chloroplast membranes of *ac-31* cells are unstacked in vivo (Fig. 17), although the thylakoids tend to lie much closer together than do the thylakoids of mixotrophic *ac-5* (compare with Fig. 12). The cell shown in Fig. 17 was grown mixotrophically; phototrophically grown cells of *ac-31*,

Table I

Chlorophyll (chl) Content and Rates of Photosynthetic Reactions of the Wild-Type, ac-5, and ac-31 Strains of C. Reinhardi.

Data are taken from single, representative experiments.

Strain	Growth condition	Total chlorophyll	Chlorophyll a:b	CO_2 fixation		DPIP photoreduction		NADP photoreduction		NADP photoreduction from DPIP-ascorbate	
		µg chlorophyll per 10^6 cells		*µmoles CO_2 fixed/hr per mg chlorophyll*	*µmoles CO_2 fixed/hr per 10^6 cells*	*µmoles photo-reduced/hr per mg chl*	*µmoles photo-reduced/hr per 10^6 cells*	*µmoles photo-reduced/hr per mg chl*	*µmoles photo-reduced/hr per 10^6 cells*	*µmoles photo-reduced/hr per mg chl*	*µmoles photo-reduced/hr per 10^6 cells*
Wild type	Mixotrophic	3.6	2.5	152	0.551	201	0.725	101	0.367	90	0.325
	Phototrophic	1.8	2.6	152	0.279	142	0.240	102	0.172	58	0.099
ac-5	Mixotrophic	0.66	3.4	230	0.151	335	0.265	134	0.106	108	0.085
	Phototrophic	0.61	3.0	230	0.140	356	0.196	128	0.070	120	0.066
*ac-31**	Mixotrophic	1.1	3.4	184	0.202	265	0.169	232	0.148	91	0.058

* Data taken from previously published experiments (21).

unlike those of *ac-5*, do not possess stacked membranes, although the tendency for membranes to "travel together" in the stroma is even more pronounced under phototrophic conditions (21).

When we isolated chloroplast fragments from *ac-31* cells by procedures identical to those followed for wild-type and *ac-5* cells, we found, to our great surprise, that the chloroplast membranes in the fragments were stacked. This is shown in Fig. 18. The in vitro stacking, moreover, does not appear to be the result of some random aggregation; rather, the thylakoids form anastomosing stacks of 2–4 thylakoids that are indistinguishable from wild-type stacks (compare Figs. 2 and 18).

In an attempt to prevent the in vitro stacking process, and thus to visualize *ac-31* membranes in their unstacked state, we tried adding glycerol slowly to a suspension of whole cells in acetate medium. After 2 hr when the glycerol concentration in the medium had reached 10%, the cells had shrunk to small compact masses; water was clearly leaving the cells in response to the presence of glycerol in the medium. When examined in thin section, these shrunken cells proved to possess stacked membranes, leading us to suspect that an increase in the ionic environment of the *ac-31* chloroplast membranes, whether brought about by removing water from whole cells or by exposing isolated fragments to high-salt buffers, might permit the membranes to stack.

On this premise we isolated *ac-31* chloroplast fragments in 0.05 M Tricine, a zwitterion that buffers effectively without introducing ions into the medium (19). The resulting preparation was found to be completely unstacked in thin section (Fig. 19); regions of even transitory contact could not be found.[2] The membranes have a somewhat

---

[2] The in vitro photosynthetic reaction rates previously published for *ac-31* (21) were undoubtedly obtained with stacked preparations; we were unaware of the in vitro stacking phenomenon at the time of this earlier study. These experiments are presently being repeated by R. P. Levine with *ac-31* fragments isolated in low-salt Tricine. The conclusions of the earlier study are not, however, discounted by the present finding, since they were based on in vivo as well as in vitro experiments.

fragmented appearance in section and stain poorly; we suspect that this reflects a poor fixation produced by the presence of Tricine in the glutaraldehyde fixation medium (13).

Freeze-etch preparations of *ac-31* membranes that have undergone in vitro stacking exhibit the four membrane faces found in wild-type and in phototrophic *ac-5* cells (Fig. 20), and these faces carry the usual distribution of particles (Figs. 9 *i* and 9 *j*). In particular, the Bs faces of in vitro stacked membranes are indistinguishable from those found within normally stacked membranes (compare Figs. 20 and 3), and they carry a substantial proportion of 160 ± 10 A particles (Fig. 9 *i*). In contrast, *ac-31* membranes that are isolated in their unstacked configuration possess only Bu and Cu faces (Fig. 21). As with the mixotrophic *ac-5* chloroplast membranes, the particle sizes on the Bu faces of this material are comparable to those found on wild-type Bu faces (compare Figs. 9 *k* and 9 *b*). Again, the particle density appears to be somewhat higher in the unstacked mutant membranes.

Unstacked Wild-Type Chloroplast Membranes

Our experience with *ac-31* indicated that membrane stacking in *C. reinhardi* is mediated by ionic interactions. To explore this point further, we isolated wild-type chloroplast fragments by our usual procedure and then washed them extensively with low-salt Tricine. As Izawa and Good had earlier found in studies of isolated spinach chloroplasts (29), such treatment leads to an almost complete unstacking of the wild-type membranes; only occasional regions of fusion remain. A freeze-etch replica of such unstacked wild-type membranes is shown in Fig. 22: only Bu and Cu membrane faces are present (compare with Fig. 3). As with the unstacked *ac-5* and *ac-31* preparations, the particle density on these Bu faces seems to be higher than that on the Bu faces found in predominantly stacked membrane preparations. Whether this difference reflects some reducton in damage suffered by the membranes during the fracturing process, or some alteration in the fracturing properties of the membranes themselves cannot be determined from this study.

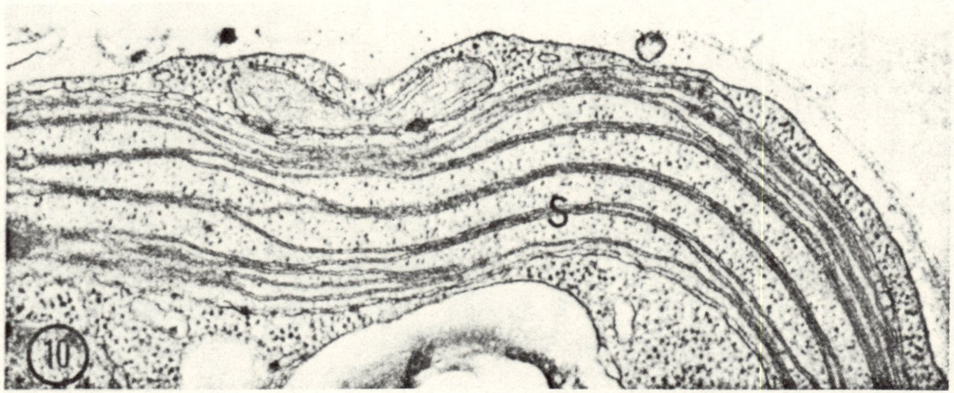

FIGURE 10 Portion of an *ac-5* cell grown phototrophically. Stacks (*S*) are primarily composed of two thylakoids each; many single discs are also present. × 48,000.

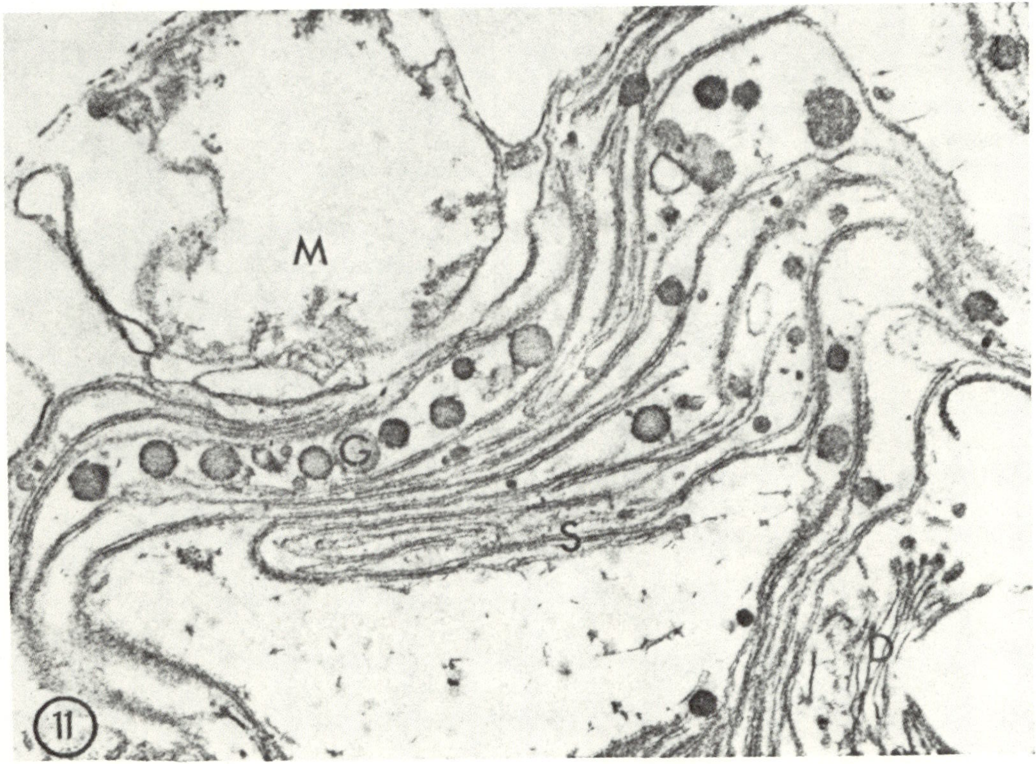

FIGURE 11 Isolated chloroplast membranes from phototrophically grown *ac-5* cells, glutaraldehyde-fixed in 27% glycerol in high salt buffer. Stacks (*S*) are composed primarily of two thylakoids each. Osmiophilic globules (*G*), a swollen mitochondrion (*M*), and a dictyosome (*D*) are also present. × 48,000.

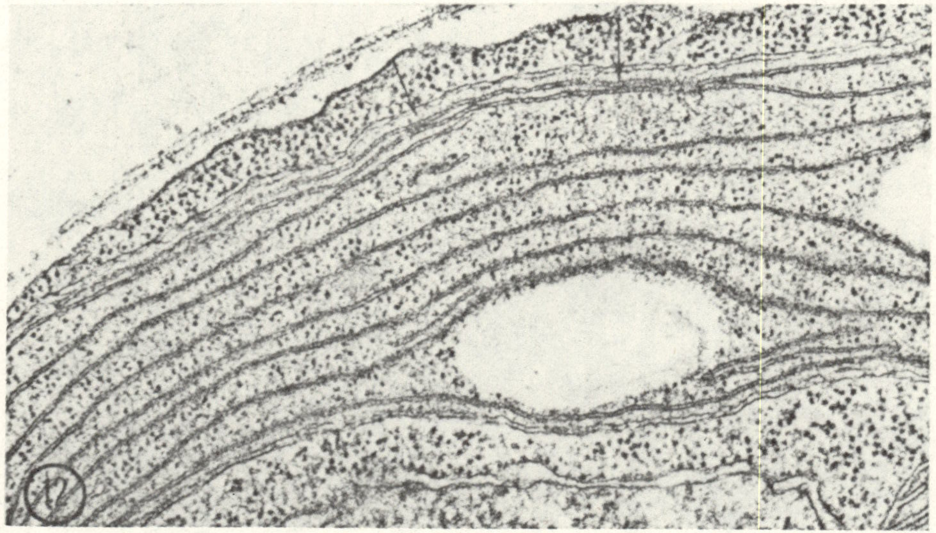

FIGURE 12 Portion of an *ac-5* cell grown mixotrophically. Single discs run an independent course through the chloroplast stroma. *Arrows* indicate limited sites where membranes approach one another, but fusion does not occur in such regions. × 48,000.

DISCUSSION

Chloroplast Membrane Faces

When biological membranes, including chloroplast membranes, are cleaved in the frozen state, the fracture plane appears to travel preferentially along the interior of the membrane in such a way that two complementary faces are exposed (6, 7, 9, 43, 45, 53, 55). In our study of the chloroplast membranes of *C. reinhardi* we have termed these two complementary faces B and C after the notation of Branton and Park (9). However, we have presented evidence to show that both faces differ in their morphology depending upon whether the membrane is found in the stacked or in the unstacked configuration. Thus, we can distinguish Bs and Cs faces in stacked regions of the chloroplast and Bu and Cu faces in unstacked regions; the two types of B faces are continuous with one another, with no clear steps or ridges between them, and the two types of C faces are similarly continuous. The relationships between fracture planes, exposed membrane faces, and particle configurations on these faces are illustrated schematically in Fig. 23, and our visualization of the organization of chloroplast thylakoids is given, again schematically, in Fig. 24.

To support our identification of the membrane faces it is important to demonstrate that Bs and Cs faces are complementary, and also that Bu and Cu faces are complementary. Comparing first the Bs and Cs faces, it is seen that the Bs face possesses a smooth background matrix marked with numerous small depressions, each 20–80 A in diameter;[3] the matrix also carries evenly distributed particles, a considerable proportion of which are 160 ± 10 A in diameter (Figs. 3, 5, and 9 *a*). The Cs face is rough and irregular, partly because it contains numerous large holes; these are located between small particles having diameters of 30–100 A each (Figs. 3, 4, 6, and 9 *c*). It thus seems

[3] As small holes and depressions are partly or completely filled with replicating material during the replication process, their apparent size is usually at least 20–30 A smaller than their real size. In thick replicas, holes with diameters of less than 60 A are often completely lost.

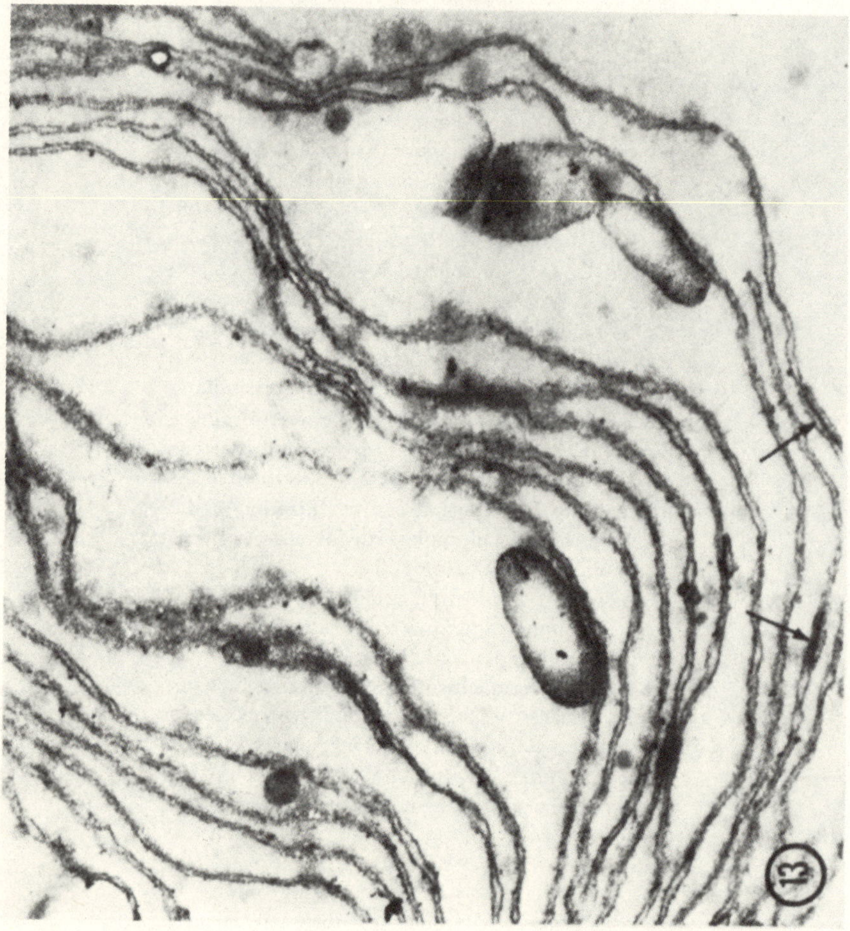
13

Figure 13 Isolated chloroplast membranes from mixotrophically grown *ac-5* cells, glutaraldehyde-fixed in 27% glycerol in high-salt buffer. Thylakoids retain their unstacked configuration. Occasional examples of membrane contact are indicated at *arrows*; as shown in Fig. 15, however, this contact does not result in a true membrane fusion. × 48,000.

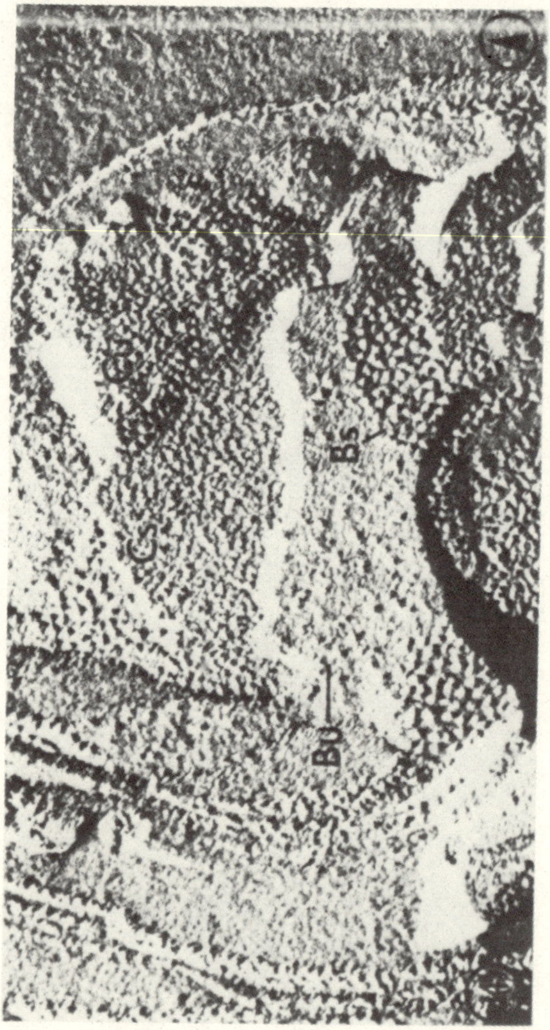
Cs
Bs
Bu

Figure 14 Freeze-etched isolated chloroplast membranes from phototrophically grown *ac-5* cells, suspended in 27% glycerol in high-salt buffer. The four faces of chloroplast membranes are indicated, and the Bs faces typical of stacked membrane regions can be recognized. The coarse shadowing of this replica covers up some of the finer details of the exposed membrane faces. Circled triangles in lower right of Figs. 14–16 indicate the angles of shadowing. × 80,000.

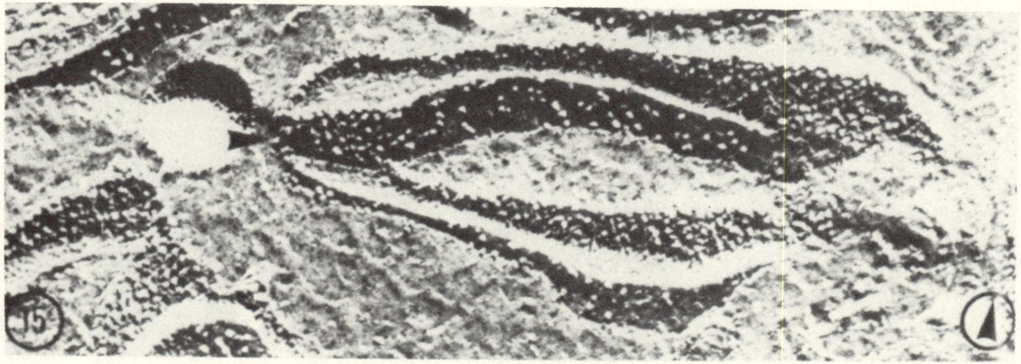

FIGURE 15 Freeze-etched, isolated chloroplast membranes from mixotrophically grown *ac-5* cells, suspended in 27% glycerol in high-salt buffer. The *arrow* indicates a "B contact" region in which two thylakoids seem to be adhering together. Such a region is readily distinguished from truly stacked regions in that it apparently carries the same complement of particles as do Bu faces, and does not carry the ordered arrays of large particles found on Bs faces. This is seen also in Fig. 9 *h*. × 80,000.

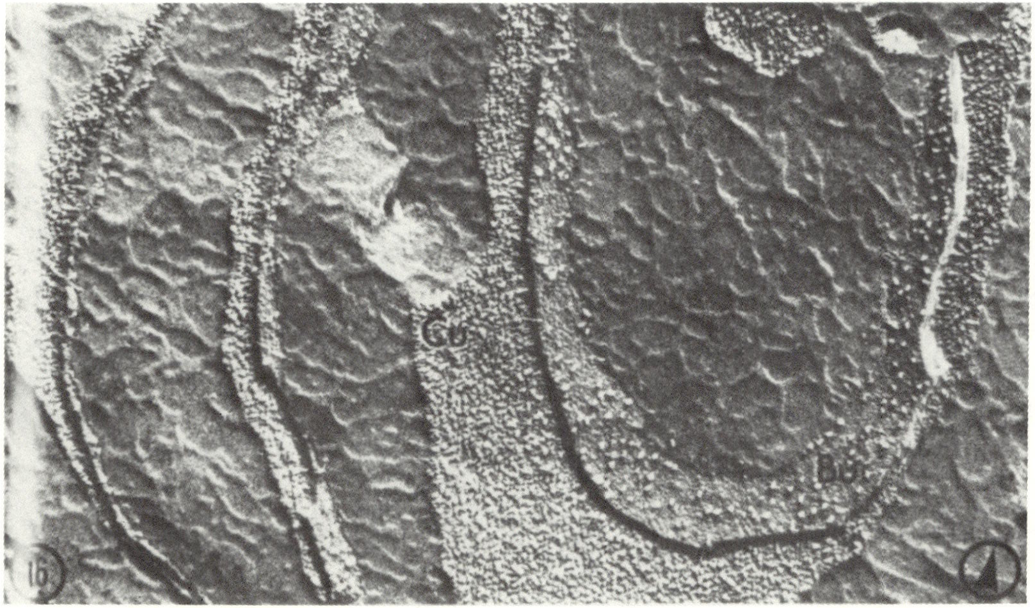

FIGURE 16 Freeze-etched, isolated chloroplast membranes from mixotrophically grown *ac-5* cells, suspended in 27% glycerol in high-salt buffer. The thylakoids are unstacked and only two different type of faces, *Bu* and *Cu*, can be distinguished. Note the higher number of particles on the Bu faces of these unstacked thylakoids compared with the Bu faces of wild-type thylakoids (Fig. 3). × 60,000.

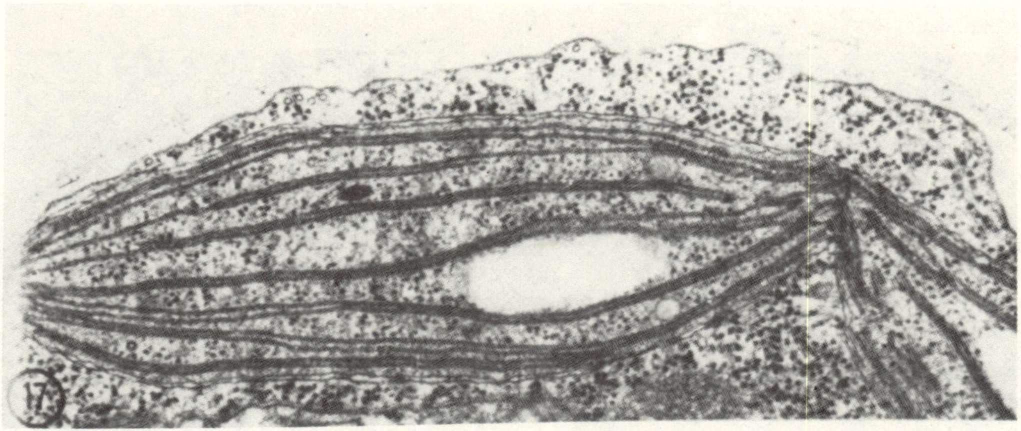

FIGURE 17 Portion of an *ac-31* cell grown mixotrophically. Single thylakoids tend to lie in closely associated pairs, but stacking does not occur between them. × 48,000.

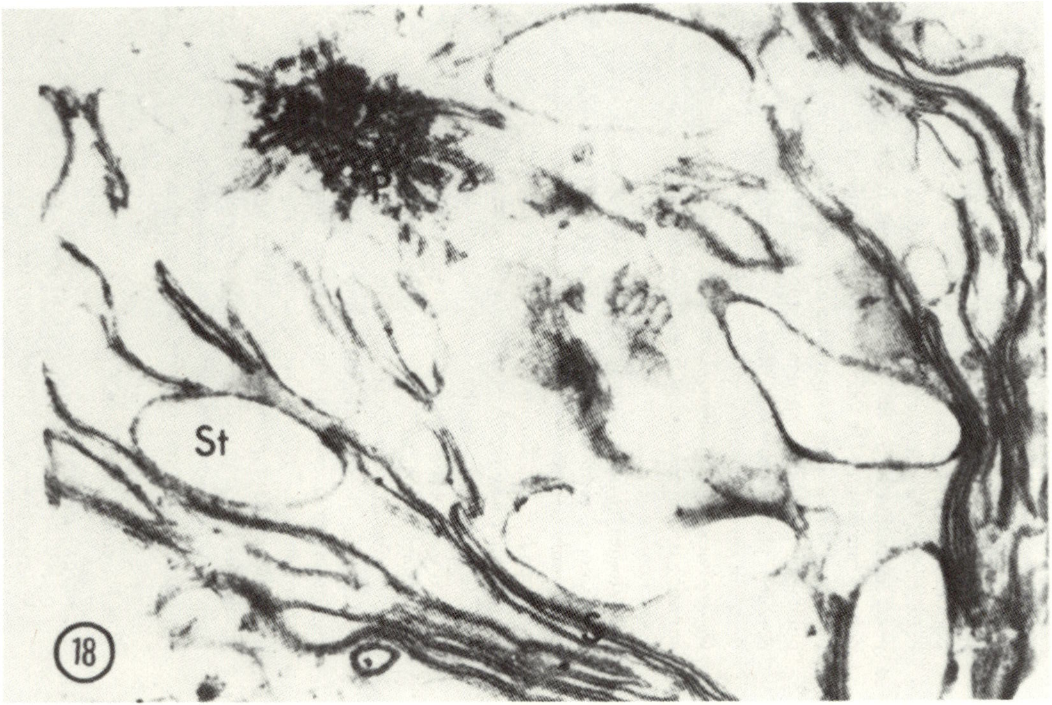

Figure 18 Isolated chloroplast membranes from *ac-31* cells, glutaraldehyde-fixed in 27% glycerol in high-salt buffer. Fusion of thylakoids has occurred in vitro, resulting in stacks (*S*) of from 2 to 5 thylakoids. A pyrenoid (*P*) and starch (*ST*) are also seen. × 48,000.

reasonable to suppose that the large particles on the Bs faces are torn out of the large holes on the Cs faces, and that the small depressions on the Bs faces are produced by the breaking away of the small particles observed on the Cs faces.

A similar "fit" can be visualized between the Bu and Cu faces. The Bu faces exhibit a background matrix that is marked by large numbers of depressions 50–120 A in diameter; these faces also carry a sparse population of variously sized particles (Figs. 3, 7, and 9 *b*). The Cu faces are richly populated with particles (Fig. 8). Where the matrix between the Cu particles can be observed, it appears smooth except for some occasional large holes. Thus Bu and Cu faces also seem to exhibit a reciprocal distribution of particles and holes. This point can be established unambiguously only by double-replica experiments (55), but our observations are certainly consistent with the notion that we are viewing complementary faces of two differentiated, but continuous, regions of the chloroplast membrane.

In studies by other investigators, the chloroplast membrane of green plants has been considered to possess only two types of inner faces (3, 9); these correspond to our Cu and Bs faces. Our Cs face has been termed the A surface by Branton and Park (9), who propose that it corresponds to the outer surface of a stacked chloroplast membrane. We cannot agree with this interpretation, one reason being that the Cs face is coplanar with the Cu face (Fig. 4) which is, by Branton and Park's interpretation as well as our own, an internal membrane face. Our Bu face is found in replicas from higher-plant chloroplasts (3, 9); other investigators have referred to it as a face which the large (Bs) particles have been pulled away during the fracturing process (see Fig. 5 of reference 9).

The Particles Found on Chloroplast Membrane Faces

The diameter of particles in freeze-etched replicas is determined, in part, by a subjective decision as to how much of the shadow of a given particle should be included in the measurement; it is also affected by such factors as pretreatment of the specimen, plastic deformation, water vapor

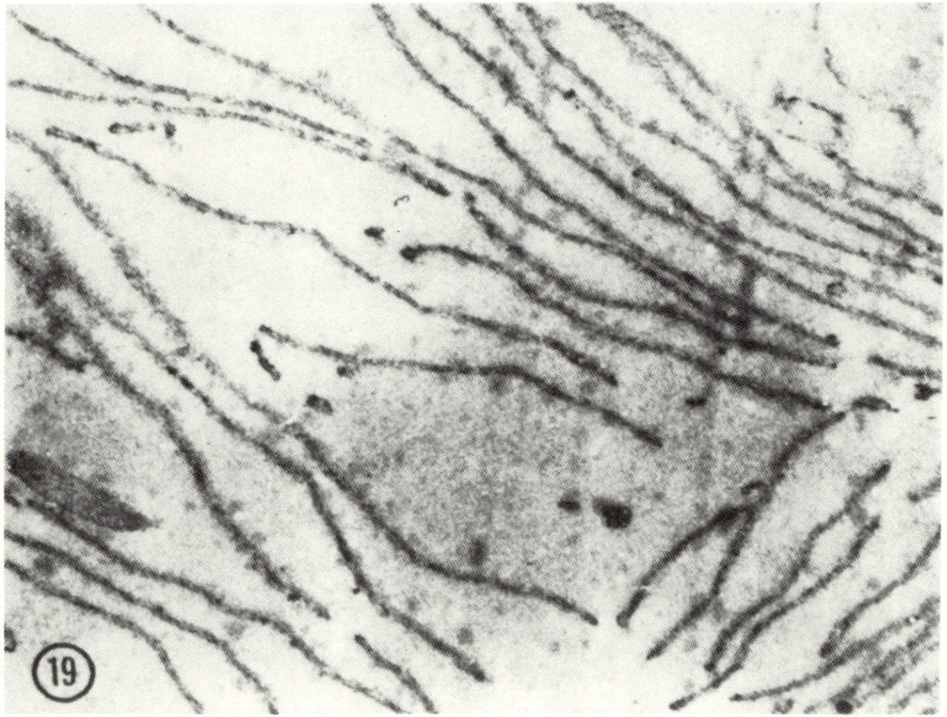

FIGURE 19 Isolated chloroplast membranes from *ac-31* cells, gluteraldehyde-fixed in 27% glycerol in Tricine buffer. Single discs remain far apart without any regions of contact. The membranes appear fragmented and stain poorly, presumably because of the glutaraldehyde fixation in Tricine. × 48,000.

sublimation and condensation, the thickness of the platinum/carbon shadowing layer, and the thickness of the carbon replicating layer. Because of these many sources of variability, we feel that a histogram of all the particle sizes found on a given membrane face (Fig. 9) is more meaningful than an average or a maximum particle size value.

As we have pointed out earlier, the particles on the Bs face are the most distinctive feature of freeze-etched chloroplast membranes. Mühlethaler et al. (39) report that this face carries a large, 120 A particle, whereas Branton and Park (9) report that the same face carries a large, 175 A particle. Our histograms (Fig. 9) indicate that this face contains a broad spectrum of particle sizes, with three maxima at 105 ± 10 A, 130 ± 10 A, and 160 ± 10 A. The average size of all of these particles lies in the 120–130 A range, which perhaps corresponds to the 120 A value of Mühlethaler et al. (39). The largest sized particles in our replicas presumably correspond to the 175 A particles of Branton and Park (9).

Our observations would indicate that to characterize chloroplast membrane faces by a single particle size is at present an oversimplification. On the other hand, we are in no position to state how many different types of particles a given face carries. It is certainly possible that some, and perhaps many, of the smaller-sized particles in a replica represent damaged fragments or dissociated subunits of larger-sized particles. For example, the fairly regular spacing of particles on the Bs face (Figs. 3, 5, 6) would suggest a more uniform population of particles on this face than the size-distribution data reveal. The peaks in the size-distribution histogram for a Bs face (Fig. 9 *a*), therefore, might reflect subunits or fragments of the 160 ± 10 A particle rather than unique species of particles. If this is the case, then a change in the conditions under which the membrane is processed might greatly increase (or decrease) the number of 160 ± 10 A particles observed; in other words, the fact that our Bs histograms are quite uniform from one preparation to the next (compare Figs. 9 *a*, *e*, and *i*) may simply reflect the uniformity of our preparative procedures.

A related kind of problem arises when comparing

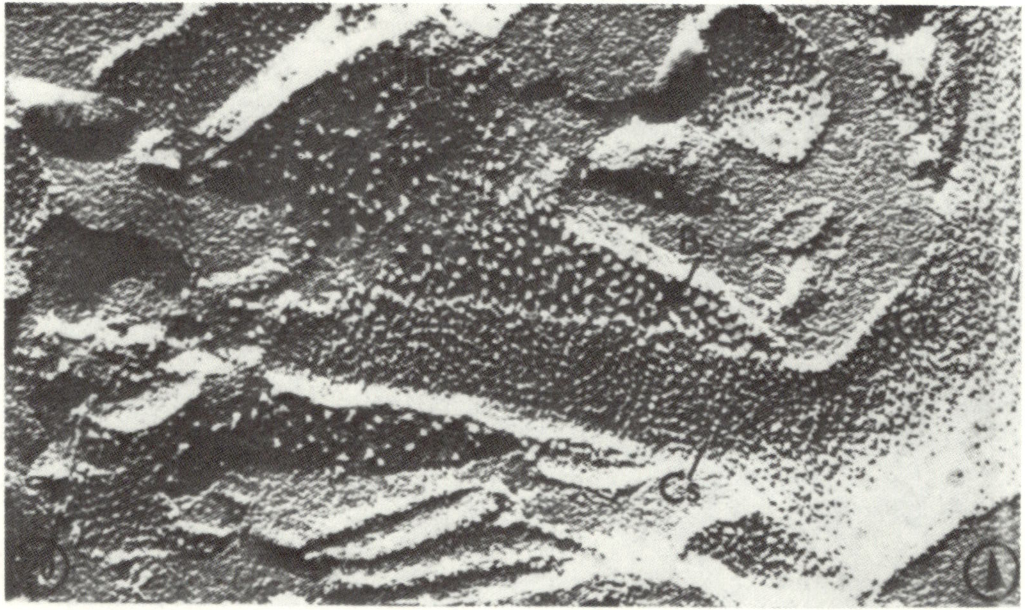

Figure 20 Freeze-etched isolated chloroplast membranes from *ac-31* cells, suspended in 27% glycerol in high-salt buffer. The membranes have stacked in vitro and exhibit faces that are indistinguishable in appearance and in their particle populations from the corresponding faces of wild-type cells (compare with Fig. 3, and see also Fig. 9, *i* and *j*). Circled triangles in lower right corner of Figs. 20–22 indicate angles of shadowing. × 80,000.

FIGURE 21 Freeze-etched, isolated chloroplast membranes from *ac-31* cells, suspended in 27% glycerol in low-salt Tricine buffer. The completely unstacked thylakoids reveal only two types of membrane faces, *Bu* and *Cu*. × 50,000.

one membrane face with another. The Bu face, for example, carries a small number of particles (some 8% of the total) with diameters of 160 ± 10 A, but we are unable to determine whether these are identical with, related to, or different from the similarly sized particles on Bs faces. Until some procedures are devised for characterizing the biochemical nature of particles observed in freeze-etch replicas, such problems will probably remain unresolved.

Chloroplast Membrane Stacking and Particle Formation

Unstacked *C. reinhardi* chloroplast membranes possess only Bu and Cu faces. This is true of naturally occurring, unstacked, wild-type membranes, as well as wild-type membranes whose stacked configuration has been experimentally disrupted by washing with low-salt Tricine; it is also true of mutant membranes that have lost the ability to stack. Conversely, the Bs and Cs faces are found only where membranes are stacked, whether this stacking has occurred within the cell or in vitro. Thus, we conclude that the Bs and Cs faces are formed as a consequence of the stacking process, and we propose that their formation reflects some conformational change that occurs within the membrane at the time of stacking. It seems reasonable to propose that, as fusion occurs between the surfaces of two membranes, unique kinds of stresses are produced within the membrane interior and that these stresses could lead to associations between membrane components that appear as particles in freeze-etch replicas.

The most conspicuous difference between the faces of stacked and unstacked chloroplast membranes is the dramatic increase of 160 ± 10 A particles on the Bs face and their fairly orderly array on that face. Thus, these particles may reflect regions of the membrane that are directly involved in stacking, which would imply that fusion takes place at discrete, closely spaced, and well-ordered intervals along the membrane surface rather than over the entire membrane surface. We have indicated this type of membrane-membrane interaction in our diagrams (Figs. 22 and 23). We stress that all of our proposals as to the

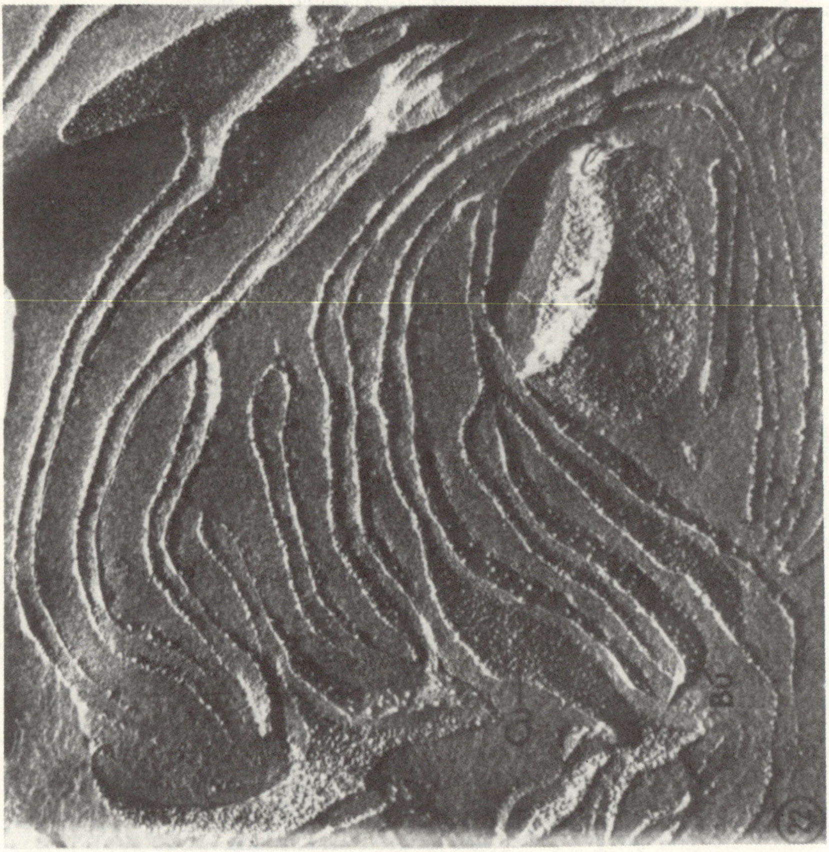

FIGURE 22 Freeze-etched, isolated chloroplast membranes from wild-type cells, isolated in high-salt buffer and then washed extensively in low-salt Tricine buffer before the addition of 27% glycerol. The thylakoid membranes have been completely unstacked by the low-salt washings. Only two types of faces, *Bu* and *Cu*, can be found in regions where the thylakoids are well separated. × 60,000.

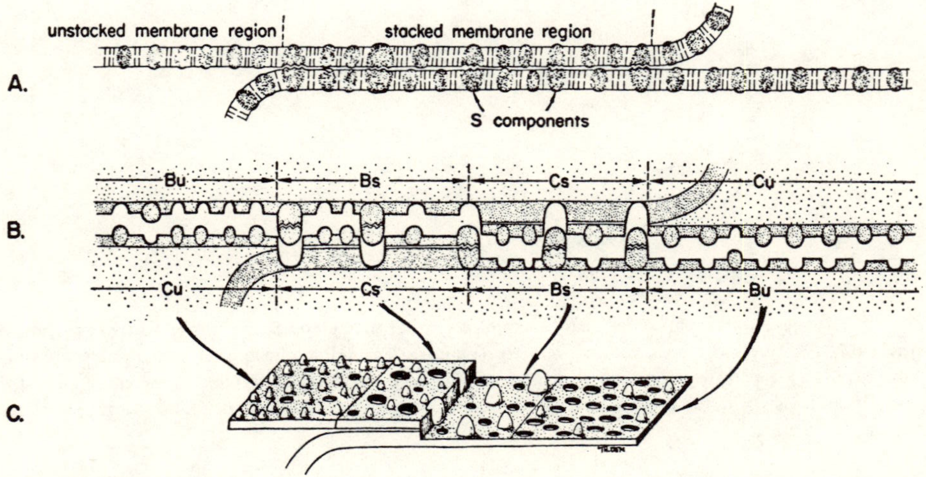

FIGURE 23 *A* Schematic drawing showing chloroplast membranes of two adjacent thylakoids that are stacked in the region defined by the two dashed lines. Elements seen in freeze-etch replicas are depicted within the membranes as stippled areas. Regions of lipid bilayer are also indicated. The *S-components* represent the suggested membrane regions that participate in the stacking process. The jagged lines indicate regions where the S components of adjacent membranes make contact. Fig. 23 *B*, Diagram indicating how the hypothetical chloroplast membrane of Fig. 23 *A* could give rise to the cleavage faces observed in freeze-etch replicas. Note the complementary features of *Bu* and *Cu*, and of *Bs* and *Cs* faces. Fig. 23 *C*, Artist's view of the faces produced during the cleaving process indicated in Fig. 23 *B*.

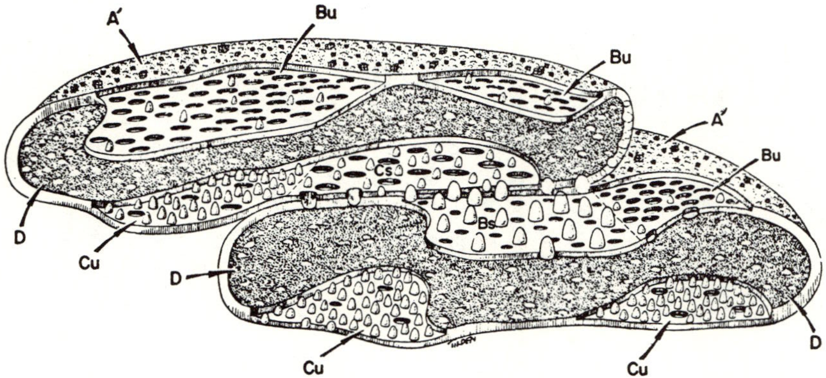

FIGURE 24 A schematic representation of two chloroplast thylakoids showing stacked and unstacked membrane regions. As revealed by freeze-etching, faces Bu and Cu and faces Bs and Cs exhibit split inner membrane faces that have complementary features. *Bu* and Cu faces are observed where the membranes are unstacked and *Bs* and *Cs* faces are observed where the membranes are stacked. The large particles on the face Bs are apparently formed in response to the stacking process. We have adopted the nomenclature and the features of the membrane surfaces marked *A′* and *D* from information provided by Arntzen et al. (3), Howell and Moudrianakis (28), and Park and Pheifhofer (42, 43).

morphological nature of chloroplast membrane stacking are speculative.

The arrays of particles that characterize stacked chloroplast membranes are specific to the chloroplast system. Thus the membrane interiors of fused myelin membranes are apparently particle free (7), and interactions between plasma membranes in junctional complexes produce freeze-etch patterns very different from those seen in chloroplast membranes (20, 49). Of particular interest is the fact that the chloroplast membranes of the mixotrophically grown *ac-5* strain of *C. reinhardi* are unable to stack in vivo; in vitro, the isolated membranes occasionally make contact, but this contact does not result in the formation of Bs and Cs particle configurations. Thus, true membrane stacking involves more than a simple apposition of chloroplast membrane surfaces; elements within the membrane interior apparently become involved in the interaction as well, and this type of interaction is evidently prevented by the *ac-5* mutation.

Stacking of photosynthetic membranes occurs in many (4, 11, 12, 15, 17, 18, 25, 27, 46, 52), but not all (10, 11, 26), photosynthetic bacteria, in all algae but the blue-green, red, and brown classes (see reference 21), and in all higher plants with the exception of those chloroplasts located in the bundle-sheath cells of certain grasses (34). We have surveyed the literature to determine the size and distribution of particles in freeze-cleaved or freeze-etched membranes from various photosynthetic organisms other than higher plants (5, 6, 14, 18, 24, 27, 48), but extensive variability in the quality of the preparations makes it very difficult to compare one micrograph with another. We have yet to encounter a replica in which distinctive arrays of large particles are unmistakably present on photosynthetic membrane faces that are not stacked. On the other hand, it is certainly possible that certain photosynthetic organisms may have evolved ways of bringing about membrane stacking that do not involve the formation of 160 ± 10 A particles. Thus, our proposal that stacking creates arrays of these particles applies, at least at present, only to the chloroplast membranes of green algae and higher plants.

The widespread occurrence of stacking among

photosynthetic organisms strongly suggests that such a membrane configuration plays some role in the photosynthetic process. No "function" for membrane stacking in photosynthesis has as yet been experimentally demonstrated, however, although several possible functions have been suggested (21). The problem is presently under investigation in the laboratory of R. P. Levine.

The 160 ± 10 A Particle and Photosynthesis

The numbers of large 160 ± 10 A particles present within chloroplast membranes are greatly enhanced as the membranes change from an unstacked to a stacked configuration, but this increase is not accompanied by any notable change in the ability of chloroplast membranes to perform photosynthetic reactions at high light intensities. Thus, *ac-5* membranes carry out PS I- and PS II-mediated reactions equally well whether stacked or unstacked (Table I), and Izawa and Good showed several years ago (29) that isolated spinach chloroplast membranes that are unstacked by low-salt Tricine are capable of substantial rates of ferricyanide photoreduction. Similarly, we are unable to relate the numbers of large particles in a membrane to the amount of chlorophyll present in the membrane.

Park and coworkers refer to the large particle as a "quantasome" or "quantasome core" (42, 43), and models of photosynthesis based on quantasome arrays have been proposed (33, 57). Recently, Arntzen et al. (3) have disrupted spinach chloroplast membranes and have obtained fractions enriched for both PS II activity and for what we have termed Bs membrane faces; they therefore suggest that the large particle of the Bs face is associated with PS II. Similarly, a membrane fraction carrying only the smaller particles found on what we term a Cu face was capable of performing only PS I–mediated reactions; thus, these investigators suggest that the smaller particles that they identify in their preparations be termed "PS I markers." Since the PS II–containing fractions are composed primarily of stacked membranes (see Fig. 6 of reference 3), and since the PS I–containing fractions are composed solely of unstacked membranes (see Fig. 10 of reference 3), we feel that the results of Arntzen et al. could be equally well interpreted by proposing that the large particle "marks" stacked membranes in their preparations, and that the smaller particle "marks" unstacked membranes in their preparations. In other words, we do not feel that any direct association between freeze-etch particles and photosynthesis has yet been demonstrated. We would like to take this opportunity to suggest that the tendency to regard the photosynthetic membrane as possessing quantized morphological structures that correspond to functional photosynthetic units be suspended until there is more conclusive evidence that such structures exist.

Factors Controlling Chloroplast Membrane Stacking in C. reinhardi

Chloroplast membrane stacking in *C. reinhardi* is conditioned, first, by the ionic environment, as has been pointed out previously for higher plants (29, 30). Thus, we are able to unstack chloroplast membranes by washing them with low-salt Tricine buffer, an observation that leads us to question the concept of stacking as a hydrophobic interaction (9). We assume that the ionic interactions required for stacking take place between membrane surfaces, and perhaps at specific sites along the membrane as we have suggested earlier in the Discussion.

Chloroplast membrane stacking in *C. reinhardi* is also controlled by at least two Mendelian genes. The *ac-5* and *ac-31* loci have been mapped to two separate linkage groups of the *C. reinhardi* genome (35), and mutations at these loci lead to quite distinct phenotypes. Chloroplast membrane stacking can occur in *ac-5* cells that are grown phototrophically, but it does not occur in mixotrophically grown cells,[4] nor can it be induced if the mixotrophic membranes are isolated in a high-salt

[4] Such a differential expressivity is known for two other gene mutations in *C. reinhardi:* the phenotypes of the *ac-20* (23, 36, 54) and the F-54 (Chua and Sato, unpublished) strains are also distinctly different under phototrophic and mixotrophic growth conditions, a phenomenon that cannot be explained at the present time.

buffer. The *ac-31* strain, on the other hand, does not exhibit stacked chloroplast membranes in whole cells under either growth condition; however, when the membranes are isolated in a high-salt buffer, they are capable of true stacking in vitro, and they can be kept apart in vitro only by isolating them in a low-salt buffer.

Two interpretations can be offered for the *ac-31* mutation. The mutation might result in some alteration of the chloroplast stroma such that the proper ionic environment for membrane stacking is not provided by the intact cells. Alternatively, the mutation might produce an alteration in some membrane component that participates in membrane stacking. As a result, the normal ionic milieu of the stroma might no longer be adequate, or appropriate, for the stacking process, whereas the high-salt buffer in the chloroplast isolation medium would be appropriate.

The buffer used to isolate stacked chloroplast membranes contains high concentrations (0.02 M) of magnesium. Since divalent cations have been frequently found to be involved in the adhesive properties of biological components, we are presently exploring the possibility that magnesium specifically mediates chloroplast membrane stacking.

It should be pointed out that Park and Pfeifhofer (43) report the presence of Bs faces in spinach membranes that have been isolated in 0.02 M Tricine. They state that A faces are rare in their preparations, implying that stacking is limited, but they do not establish by thin sectioning that their membrane fragments are totally unstacked. Anderson and Vernon (1), repeating the Tricine procedures of Izawa and Good (29), obtain only partially unstacked spinach preparations, and Kahn and von Wettstein (32) observe stacking in spinach chloroplast fragments that are suspended in distilled water. We find that, in *C. reinhardi*, extensive washing in five changes of low-salt Tricine buffer still leaves some intact stacks in wild-type membrane preparations. Taken together, these results indicate that rigorous criteria must be used to determine that a given preparation is unstacked, particularly when one is attempting to pull apart membranes that are initially tightly fused.

The Differentiation of the Chloroplast Membrane

Izawa and Good (29) have reported that when they add salts back to isolated spinach chloroplasts that have been unstacked in a low-salt Tricine buffer, normal looking grana often assemble in vitro. We have made a similar observation with *C. reinhardi*: when *ac-31* membranes are provided with a high-salt environment in vitro, the normal *C. reinhardi* pattern of membrane association occurs, and not some massive, nonspecific aggregation of membranes. Since this pattern is formed by isolated, washed membranes, we conclude that a specific basis for the patterning is built into the structure of the membranes themselves. In other words, it appears that certain regions of a chloroplast membrane are capable of stacking, and that adjacent regions are not. We have not been able to detect any obvious manifestation of this patterning in freeze-etch replicas; thus the Bu and Cu faces of wild-type and *ac-31* membranes that have been isolated in low-salt Tricine do not reveal any patterning of particle sizes or densities that we might relate to potential regions of stacking and unstacking. This does not seem surprising, however, for we presume that if a morphological manifestation of chloroplast membrane differentiation exists at all, it should be found on the membrane surface and not in the membrane interior.

We thank Professors R. P. Levine and K. R. Porter for providing the laboratory facilities in which this investigation was carried out. We are grateful for the helpful advice and considerable interest given to this project by Professor Levine, and for the expert technical assistance of Mr. Henry V. Webber III.

Dr. Goodenough was supported by a postdoctoral fellowship (GM-24036) from the National Institutes of Health, and Dr. Staehelin was supported by a Health Sciences Advancement Award Grant 5-SO4-FR06084 to the University of Colorado. The research was also supported by a grant (GB-5005-x) from the National Science Foundation to R. P. Levine.

Addendum. After this manuscript had been submitted for publication, two articles that we had not previously seen were brought to our attention: Remy, R. 1969. Etude de la structure des lamelles chloroplastiques de *Vicia faba* L. par la technique du cryodécapage. *C. R. Acad. Sci.* **268**:3057, and Phung Nhu Hung, S., A. Lacourly, and C. Sarda. 1970. Etude de l'evolution en chloroplastes des plastes etiolés d'orge. I. Structure examinée par cryodécapage. *Z. Pflanzenphysiol.* **62**:1. These authors have observed that membrane faces identical to our Bs faces occur only in stacked regions of bean and barley chloroplasts, respectively. We acknowledge the priority of these authors in making this observation. It appears that, at least on this point, our interpretation of chloroplast membrane structure applies not only to *Chlamydomonas* but to higher plants as well.

REFERENCES

1. Anderson, J. M., and L. P. Vernon. 1967. Digitonin incubation of spinach chloroplasts in tris (hydroxymethyl) methyl glycine solutions of various ionic strengths. *Biochim. Biophys. Acta.* **143**:363.
2. Arnon, D. 1949. Copper enzymes in isolated chloroplasts. Polyphenol oxidases in *Beta vulgaris*. *Plant Physiol.* **24**:1.
3. Arntzen, C. J., R. A. Dilley, and F. L. Crane. 1969. A comparison of chloroplast membrane surfaces visualized by freeze-etch and negative staining techniques; and ultrastructural characterization of membrane fractions obtained from digitonin-treated spinach chloroplasts. *J. Cell Biol.* **43**:16.
4. Boatman, E. S., and H. C. Douglas. 1961. Fine structure of the photosynthetic bacterium *Rhodomicrobium vanniellii*. *J. Biochem. Biophys. Cytol.* **11**:469.
5. Bourdu, R., and M. Lefort. 1967. Structure fine, observée en cryodécapage, des lamelles photosynthétiques des Cyanophycées endosymbiotiques: *Glaucocystis nostochinearum* Itzigs et *Cyanophora paradoxa* Korschikoff. *C. R. Acad. Sci.* **265**:37.
6. Branton, D. 1966. Fracture faces of frozen membranes. *Proc. Nat. Acad. Sci. U.S.A.* **55**:1048.
7. Branton, D. 1967. Fracture faces of frozen myelin. *Exp. Cell Res.* **45**:703.
8. Branton, D. 1969. Membrane structure. *Ann. Rev. Plant Physiol.* **20**:209.
9. Branton, D., and R. B. Park. 1967. Subunits in chloroplast lamellae. *J. Ultrastruct. Res.*

19:283.

10. Cohen-Bazire, G., N. Pfennig, and R. Kunisawa. 1964. The fine structure of green bacteria. *J. Cell Biol.* **22**:207.
11. Cohen-Bazire, G., and W. R. Sistrom. 1966. The procaryotic photosynthetic apparatus. *In* The Chlorophylls. L. P. Vernon and G. R. Seely, editors. Academic Press Inc., New York. 313.
12. Conti, S. F., and P. Hirsch. 1965. Biology of budding bacteria. III. Fine structure of *Rhodomicrobium* and *Hyphomicrobium* spp. *J. Bacteriol.* **89**:503.
13. De Petrocellis, B., P. Siekevitz, and G. E Palade. 1970. Changes in chemical composition of thylakoid membranes during greening of the *y*-l mutant of *Chlamydomonas reinhardi. J. Cell Biol.* **44**:618.
14. Dilley, R. A. 1969. Observations on the structure of *Rhodopseudomonas viridis* as shown by freeze etch. *In* Progress in Photosynthesis Research. H. Metzner, editor. Tübingen Press. Tübingen, Germany. 197.
15. Drews, G., and P. Giesbrecht. 1965. Die thylakoidstrukturen von *Rhodopseudomonas spec. Arch. Mikrobiol.* **52**:242.
16. Farquhar, M. G., and G. E. Palade. 1963. Junctional complexes in various epithelia. *J. Cell Biol.* **17**:375.
17. Gibbs, S. P., W. R. Sistrom, and P. B. Worden. 1965. The photosynthetic apparatus of *Rhodospirillum molischianum. J. Cell Biol.* **26**:395.
18. Giesbrecht, P., and G. Drews. 1966. Uber die organization and die makromolekulare Architektur der Thylakoide "lebender" Bakterien. *Arch. Mikrobiol.* **54**:297.
19. Good, N. E. 1962. Uncoupling of the Hill reaction from photophosphorylation by anions. *Arch. Biochem. Biophys.* **96**:653.
20. Goodenough, D. A., and J. Revel. 1970. A fine structural analysis of intercellular junctions in the mouse liver. *J. Cell Biol.* **45**:272.
21. Goodenough, U. W., J. J. Armstrong, and R. P. Levine. 1969. Photosynthetic properties of *ac-31*, a mutant strain of *Chlamydomonas reinhardi* devoid of chloroplast membrane stacking. *Plant Physiol.* **44**:1001.
22. Goodenough, U. W., and R. P. Levine. 1969. Chloroplast ultrastructure in mutant strains of *Chlamydomonas reinhardi* lacking components of the photosynthetic apparatus. *Plant Physiol.* **44**:990.

23. Goodenough, U. W., and R. P. Levine. 1970. Chloroplast structure and function in *ac-20*, a mutant strain of *Chlamydomonas reinhardi*. III. Chloroplast ribosomes and membrane organization. *J. Cell Biol.* **44**:547.

24. Guérin-Dumartrait, E. 1968. Etude, en cryodécapage, de la morphologie des surfaces lamellaires chloroplastiques de *Chlorella Pyrenoidosa*, en cultures synchrones. *Planta.* **80**:96.

25. Hickman, D. D., and A. W. Frenkel. 1965 *a*. Observations on the structure of *Rhodospirillum molischianum*. *J. Cell Biol.* **25**:261.

26. Hickman, D. D., and A. W. Frenkel. 1965 *b*. Observations on the structure of *Rhodospirillum rubrum*. *J. Cell Biol.* **25**:279.

27. Holt, S. C., H. G. Trüper, and B. J. Takács. 1968. Fine structure of *Ectothiorhodospira mobilis* strain 8113 thylakoids: chemical fixation and freeze-etching studies. *Arch. Mikrobiol.* **62**:111.

28. Howell, S. H., and E. N. Moudrianakis. 1967. Function of the "quantasome" in photosynthesis: structure and properties of membrane-bound particle active in the dark reactions of photophosphorylation. *Proc. Nat. Acad. Sci. U.S.A.* **58**:1261.

29. Izawa, S., and N. E. Good. 1966. Effect of salts and electron transport on the conformation of isolated chloroplasts. II. Electron microscopy. *Plant Physiol.* **41**:544.

30. Jacobi, G., and H. Lehman. 1969. Photochemical activities of chloroplast fragments. *In* Progress in photosynthesis research. H. Metzner, editor. Tübingen Press, Tübingen, Germany. 159.

31. Johnson, U. G., and K. R. Porter. 1968. Fine structure of cell division in *Chlamydomonas reinhardi*. Basal bodies and microtubules. *J. Cell Biol.* **38**:403.

32. Kahn, A., and D. von Wettstein. 1961. Macromolecular physiology of plastids. II. Structure of isolated spinach chloroplasts. *J. Ultrastruct. Res.* **5**:557.

33. Kelly, J., and K. Sauer. 1965. Action spectrum and quantum requirements for the photoreduction of cytochrome c with spinach chloroplasts. *Biochemistry.* **4**:2798.

34. Laetsch, W. M., and I. Price. 1969. Development of the dimorphic chloroplasts of sugar crane. *Amer. J. Bot.* **56**:77.

35. Levine, R. P., and U. W. Goodenough. 1970. The genetics of photosynthesis and of the chloroplast in *Chlamydomonas reinhardi*. *Ann. Rev.*

Genet. **4**:397.

36. LEVINE, R. P., and A. PASZEWSKI. 1970. Chloroplast structure and function in *ac-20*, a mutant strain of *Chlamydomonas reinhardi*. II. Photosynthetic electron transport. *J. Cell Biol.* **44**:540.

AUTHOR INDEX

KEY-WORD TITLE INDEX

RA recent articles & research in progress **RIP**

Guide to Current Research

The research summaries appearing in the following section were obtained through a search of the Smithsonian Science Information Exchange data base conducted in January, 1974.

The Exchange annually registers 85,000 to 100,000 notices of current research projects covering a wide range of disciplines and sources of support. SSIE endeavors to retain up to two full years of current research information in its active file. The selection of summaries appearing in this section does not represent the complete SSIE collection of information on this topic, but, rather, has been specifically tailored to reflect the scientific content of this particular volume. A limited number of summaries may have been omitted because clearance for publication by the supporting agency or organization was not received prior to the publication date.

SSIE is the only, single source for information on ongoing and recently terminated research in all areas of the life, physical, behavioral, social and engineering sciences. The SSIE file is updated daily by a professional staff of scientists utilizing a comprehensive and flexible system of hierarchical indexing. Retrieval of subject information is conducted by these same specialists, using computer-connected, video display terminals which allow instant access to the entire data base and on-line refinement of search strategies. SSIE offers an information service unequalled anywhere: comprehensive and vital information on who is conducting what research where and under whose support.

More current information, and in some cases expanded coverage, on the topic considered in this volume is available directly from SSIE. This information is offered at modest cost in the form of custom searches of the SSIE file designed specifically to meet the user's need or as an update of the subject search in this section. For more information on SSIE, contact MSS or write directly to the Smithsonian Science Information Exchange, 1730 M Street, N.W., Washington, D.C. 20036. Subject search or updated package requirements may be discussed with SSIE scientists by calling the Exchange at (202) 381-5511.

PHOTOSYNTHESIS OF CLADCPHORA IN GREEN BAY,
M.S. ADAMS, Univ. of Wisconsin, Inst. of Environmental Studies, Madison, Wisconsin 53706

Objectives: Objectives were to measure net photosynthesis of Cladophora glomerata in Green Bay in relaticn to several growth parameters: temperature, light, and nutrients in the water and plant tissues. Bay circulation and distance from the Fox River, a source of nutrients, were considered in making three site selections. The study was directed toward increasing the understanding of seasonal and diurnal patterns of net photosynthesis of Cladophora glomerata under field conditions. We consider data from the study prerequisite to the development of a productivity model for C. glomerata in Green Bay.

How information will be applied: The information will be used to develop a net photosynthesis and productivity model for C. glomerata. Laboratory studies are in progress to determine appropriate rate functions under controlled conditions. Later field data will be used in model validation. The model will be compared with one being developed for a Oedogonium in a different body of water (Lake Wingra, Wisconsin), the objective being to develop a generalized model of periphyton productivity. Models such as the ones being developed will be useful tocls in the eventual control of such algal nuisance species, since the models will be based on data taken from or simulating actual field conditions.

Accomplishments during the past twelve months: Seasonal differences in productivity were related to seasonal temperature changes. Site differences in productivity were related to nutrient levels in the plant tissues. Under statistically similar temperature and irradiance conditions, higher rates of net photosynthesis occurred at the site which receives the greatest concentration of pollution from the Fox River. Nitrogen, phosphorus, calcium, strontium and sodium were highest in the plants from the high production site.

Contrary to other report, Cladophora at the sites we studied made relatively efficient use of low illumination.

For additional information pertaining to this project contact Dr. Robert A. Ragotzkie, Sea Grant Program Coordinator, University of Wisconsin, Madison, Wisconsin 53706.

SUPPORTED BY:
U.S. Dept. of Commerce - N.O.A.A.

H ION EXTRUSION AS THE PRIMARY ELECTROGENIC MECHANISM IN THE NITELLA MEMBRANE,

C.E. BARR, State University of New York, School of Natural Sciences, Brockport, New York 14420

The present grant is a continuation of research now supported by the National Science Foundation and deals with active H ion extrusion as the primary electrogenic mechanism in the giant cells of the alga, Nitella clavata. The proposed work focuses on four main points: a) Artificially controlling the H ion extrusion rate by injection of H ion into cells, b) dependence of the membrane potential on H ion extrusion, c) dependence of H ion extrusion on light and O2, and d) relation of H ion extrusion to internal pH.

The presence of a high rate of H ion extrusion electrically balanced by an equivalent passive inward current of H ion in Nitella appears to be fairly well established. The question of whether or not the resting pctential is essentially by H ion fluxes has been a difficult one to answer since quantitative data on the unidirectional fluxes are intrinsically difficult tc obtain.

The present plan is to bring abcut an outward net H ion flux by injecting H ion into the cell. The hypothesis here is that a slight lowering of the internal pH will induce a sizable increase in H ion extrusion which will result in a hyperpolarization of the membrane.

SUPPORTED BY:
U.S. Natl. Science Foundaticn

METABOLIC REGULATION OF PHOTOSYNTHESIS IN CHLORELLA PYRENOIDOSA,

J.A. BASSHAM, Univ. of California, Lawrence Berkeley Laboratory, Berkeley, Califcrnia 94720

Unicellular algae, Chlcrella pyrenoidosa, are grown in continuous culture apparatus, harvested, centrifuged, and placed in steady-state apparatus for kinetic studies of photosynthesis in the presence of tracers, 14CO2 and 32P labeled phosphate. After a period of photosynthesis, some change in environment (such as turning off the light) or addition cf a chemical (such as a fatty acid) or nutrient (such as ammonium ion) is made. Samples of the algae culture are taken at frequent intervals before and after the change. These samples are killed, and the radioactive compcunds are separated by two-dimensional paper chrcmatcgraphy and detected by radioautography. The amount of each label in each compound at all times before and after the change is determined. From the levels of radioactive label in the compounds, levels of metabolites, rates of flow of material, regulation of reactions in vivo, relations between chloroplasts and cytoplasm, relations between light and dark metabolism, and many other metabolic properties are determined. This method provides valuable information about the control of metabolism in vivo which cannot be obtained in any other way.

Results: Pool sizes of metabolites and rates of through-flow determined, sites of regulation within the photosynthetic carbon reduction cycle have been identified, regulatory sites on biosynthetic paths leading from the cycle identified, synthesis of protein from free amino acid pools demonstrated, quantitative importance of the cyclic reductive pentose phosphate cycle of photosynthesis demonstrated, and many others.

SUPPORTED BY:
U.S. Atomic Energy Commission

INVESTIGATION OF TRITIUM METABOLIC FATE IN CHLORELLA PYRENOIDOSA,

J.A. BASSHAM, Univ. of California, Lawrence Berkeley Laboratory, Berkeley, California 94720

Chlorella pyrenoidosa will be grown in a closed system at constant density, for several generations, with water enriched with T20 and with recirculating 14CO2. The algae will be harvested, killed in methanol, the exchangeable T removed, and the material from the cells extracted and/or hydrolyzed.

Initially, fats, proteins, water soluble compounds, and perhaps nucleic acids will be separated. The T/C14 ratios of these compound classes will be determined. Based on numbers of H and C atoms, and number of exchangeable H positions, calculations show whether there is tritium enrichment in any compound classes. Where T enrichment is found, it warrants extensive further study into the nature of the biochemical reactions responsible, and the identity of the T-enriched compounds. The comparative T/C ratios in green algae and blue-green algae provide information about differences in metabolism between the different kinds of photosynthetic cells.

Results: A wide variety of T/C ratios were found ranging from 0.4 to 1.6. In particular, T/C ratios were higher for glutamate, aspartate, citrate, and other compounds related to the tricarboxylic acid cycle than for sugar phosphates and sugar (in green algae).

SUPPORTED BY:
U.S. Atomic Energy Commission

BIOGENESIS OF CHLOROPLAST MEMBRANES OF CHLAMYDOMONAS REINHARDI,

D.P. BECK, Harvard University, School of Arts, Cambridge, Massachusetts 02138

The relative times of synthesis of specific chloroplast membrane proteins and lipids are being investigated in synchronously growing Chlamydomonas reinhardi. Cells of C. reinhardi can be induced to divide synchronously in response to a light-dark cycle. Under this condition, many cellular constituents are known to increase at specific times during the growth cycle. In these studies, a double-labelling technique is employed to determine when specific proteins and lipids are synthesized and incorporated into the membranes of the chloroplast. Proteins are separated by disc gel electrophoresis in SDS-urea gels and the radiometric data are analyzed statistically by computer. Membrane lipids are separated by thin layer chromatography and measured both radiometrically and chemically. Studies are planned which will determine the effects of antibiotics on the lipid and protein synthesis and incorporation into membranes. The aim of the program described above is to elucidate the relationship between membrane lipid and protein synthesis, and it should answer some questions concerning the mechanism of chloroplast membrane biogenesis.

SUPPORTED BY:
Helen Hay Whitney Found. - New York, N.Y.

MUTANT STUDIES ON THE PHOTOSYSTEM II PORTION OF THE PHOTOSYNTHETIC PROCESS,

N.I. BISHOP, Oregon State University, School of Science, Corvallis, Oregon 97331

The principle objective of the proposed work is to gain further information concerning the mechanism of oxygen production (or Photosystem II) by employing selected mutants of the green algae Scenedesmus obliquus and Chlorella pyrenoidosa. Approximately 30 mutant strains of these algae have been isolated which are blocked in the photosystem II portion of the photosynthetic mechanism. Previous studies on some of these mutants suggest that the cause of the mutation may reside in alteration of components of the membranes which are essential for this activity; these findings including alteration of chlorophyll, carotenoid, plastoquinone, cytochrome b-559 and C-550 contents. It is proposed to isolate the individual photosystem I and photosystem II particles of the normal and mutant forms of the algae and to examine their lamellar proteins and tryptic-peptides in order to determine if fundamental deletions have occurred. Comparative studies will be made on mutants that are known to lack PS-I, mutants that are deficient in carotenoids, mutants that possess abnormal chloroplast ultrastructure and mutants form normal chloroplast membranes only when explosed to light. Comparative studies will be made, where feasible, on the types of cytochrome b-559 present in the different mutant types.

Results: In studies on the mutant phenotypes that lack PS-II activity it has been the general observation that these forms all have high fluorescence, altered low-temperature fluorescence spectra, decreased amounts of various components associated with PS-II, and also lack the characteristic delayed light emission and E.P.R. signal of this photosystem. We have found that it is possible to isolate and purify the pigment-protein complex having the characteristics of a similar isolate from spinach chloroplasts. This component shows electrophoretic characteristics that make it amenable for further subdivision and analysis; this procedure would be extended to the mutants.

SUPPORTED BY:
U.S. Atomic Energy Commission

MECHANISM OF OXYGEN PRODUCTION IN PHOTOSYNTHESIS,

N.I. BISHOP, Oregon State University, Agricultural Experiment Sta., Corvallis, Oregon 97331

Objective: Gain further understanding of the mechanism of oxygen production by green plants. Information will be sought on the make-up of the electron transport system associated with the reducing and oxidizing sides of photosystem II, that portion of the mechanism responsible for water photolysis and oxygen production. Through mutant induction and analyses further information about the specific roles of the plastoquinones, b-type cytochromes, and manganese will be sought. Special attention will be directed to the role of a new heat-stable cofactor of chloroplast in the mechanism of photosystem II. Through mutant studies, the importance of the accessory pigments, notably the carotenoids, in the construction of photosynthetically competent chloroplast will be examined.

Approach: Most of the objectives of the project will be studied with the aid of photosynthetic mutants derived from a number of single celled green algae. Many of the desired mutants have been isolated and are under study. With newly derived methods of induction, isolation, and selection of mutants, more sophisticated classes of mutant phenotypes for this project will be at hand. Biochemical purification and analysis on the various electron transport components of chloroplasts, plus studies on whole cell reactions including fluorescence analysis, will expedite the development of this project.

Progress: Research activities included: 1) Studies on the development of photosystem II activity in Scenedesmus mutants blocked at specific points in carotenoid biosynthesis. 2) Studies on the greening process of pigment mutants. 3) The effect of certain antibiotics which alter the transcriptional and translational properties of chloroplast DNA and RNA. 4) The loss of an in vivo cyclic photophorylation in mutants shown to be deficient in cytochrome f(552). 5) Partial restoration of Hill reaction activity in chloroplast particles of PS-II mutants by a heat stable protein factor isolated from spinach. Detailed studies using antibiotics on carotene mutants of Scenedesmus have shown that the production of fully active chloroplast membranes is dependent upon: 1) the nature of the carotenoid; 2) the ability to synthesize chlorophyll and 3) the capacity for normal synthesis of chloroplast membranes.

SUPPORTED BY:
Oregon State Government - Salem

PHOTOSYNTHETIC ELECTRON TRANSPORT,
F.L. CRANE, Purdue University, School of Science, Lafayette, Indiana 47907

The pathways of electron transport associated with energy coupling processes in chloroplasts will be determined. Selective removal of depletion of components will be accomplished by mutation, physiological stress or fragmentation of chloroplasts as to determine the order and organization of the electron transport system. Emphasis will be placed on the protein components of grana photosystem II, grana photosystem I and stroma photosystem I, respectively. Experimental systems will include chloroplasts from mineral- deficient maize and mutants of Chlorella.

SUPPORTED BY:
U.S. Natl. Science Foundation

INDUCTION OF CELL DIVISION AND NUCLEAR REPLICATION IN PLANTS BY 2,4-D,
M.E. DUYSEN, North Dakota State University, Agricultural Experiment Sta., Fargo, North Dakota 58103

Objective: Delimit the degree of resistance of cells to 2,4-D, ascertain the susceptibility of cells at different stages of the cell cycle, and determine the mechanism of action of 2,4-D in the induction of nuclear and nucleic acid replication.

Approach: Chlorella pyrenoidosa will be grown axenically in chemostats either heterotrophically, autotrophically, or mixotrophically. 2,4-D will be added to the growth medium at different concentrations and the growth rate will be monitored. Cells treated with low concentrations of 2,4-D will be moved to higher concentrations and monitored. Synchronous cultures will be treated with 2,4-D at different times in the life cycle. Growth and DNA content will be monitored. Chromatin isolated from 2,4-D treated cells will be monitored as to its action on in vitro DNA synthesis.

SUPPORTED BY:
U.S. Dept. of Agriculture - C.S.R.S.

REGULATION OF PROTEIN FORMATION IN HIGHER PLANT CELLS AND ITS SIGNIFICANCE IN GROWTH REGULATION,
P. FILNER, Michigan State University, School of Agriculture, East Lansing, Michigan 48823

Eukaryotic cells contain pipelike structures called microtubules. They are made of protein which is very similar regardless of origin. Microtubules function in mitosis, cell movement, intracellular granule movement and cell shape determination. There are several reports in the literature which indicate that cAMP is involved in regulation of microtubules in mammals. We are examining this possibility in green plants. Our pilot study has been concerned with microtubules of the flagella of Chlamydomonas reinhardi. We have found that this organism contains cAMP and a cyclic nucleotide phosphodiesterase which can be inhibited by methylxanthines. When incubated in aminophylline, the cAMP level in C. reinhardtii increased more than 10-fold. Aminophylline inhibits flagellar movement and flagellar regeneration, as does caffeine, but dibutyryl cAMP does not. However, in mutants selected for impaired motility due to structural abnormalities in their flagella, flagellar regeneration was subject to synergistic inhibition by aminophylline plus dibutyryl cAMP. These results show that cAMP affects a flagellar component. We shall determine if it is microtubule protein.

SUPPORTED BY:
U.S. Atomic Energy Commission

A STUDY OF NATURALLY OCCURRING ALGICIDES PRODUCED BY FRESH WATER ALGAE,

D.O. HARRIS, Univ. of Kentucky, Graduate School, Lexington, Kentucky 40506

The purpose of this study is to examine the biological and chemical properties of algicides produced by fresh-water green algae (Volvocales). Past experiments by the author have demonstrated that a complex system of growth inhibition is in operation in this group. Pandorina morum, as the producer of the strongest inhibitor in the group, and Volvox tertius, as the most sensitive to these inhibitors will be used as a model system to study naturally occurring algicides. Preliminary investigations of this model system have yielded the following information: (a) the inhibitor is relatively stable at high temperatures, (b) the inhibitor appears on the 12th day after inoculation of Pandorina into fresh medium, (c) the substance can be diluted several times with the retention of at least partial activity, (d) the material moves slowly through a dialysis membrane, (e) it is unaffected by proteolytic enzymes and (f) results with Sephadex gel filtration suggests a molecular weight of less than 1500.

The study here will be three-fold in direction: (1) a biochemical investigation into how and where the substance operates in the cell to stop growth. Preliminary investigation suggests that the photosynthetic system is adversely affected in that only one hour exposure to the inhibitor severly curtails oxygen evolution. (2) A complete chemical characterization of the substance. (3) Field and laboratory studies to determine the effectiveness of these substances in controlling nuisance algae growth under field or natural conditions. Perhaps this line of investigation will eventually provide clues as to a method of controlling algae in water supplies, etc.

SUPPORTED BY:
U.S. Dept. of Interior - O. Wtr. Res. Rch.

THE EFFECT OF CADMIUM AND ZINC ON FRESHWATER ALGAE

B.A. HART, Univ. of Vermont, School of Medicine, Burlington, Vermont 05401

It is the purpose of our proposed research to determine possible effects of the heavy metals cadmium and zinc on selected species of freshwater algae, including strains isolated from the plankton. Initial studies will be done using those algae (Anacystis nidulans and Anabaena variabilis) which have been grown in mass culture in our laboratory for the past year as part of another project. Some plankton algae (Cyclotella sp., Asterionella formosa, and Chlorella sp.) are currently available in unialgal culture and will be investigated as soon as suitable methods are worked out for large scale culture. It is expected that other strains will be added to the study on the basis of their response to metals in mixed cultures.

Three topics which we intend to include in our investigation are listed as follows: 1. Effect of the metals on mixed phytoplankton cultures and the isolation of sensitive and tolerant strains. 2. Effect of the metals on metabolic processes including growth rate, oxygen evolution, autotrophic CO_2 fixation, and pigment content. 3. The uptake and efflux of heavy metals from algal cells and identification of the cellular site of metal deposition.

SUPPORTED BY:
U.S. Dept. of Interior - O. Wtr. Res. Rch.

BIOGENESIS OF CELLULAR MEMBRANES,

J.K. HOOBER, Temple University, School of Medicine, Philadelphia, Pennsylvania 19122

During greening of etiolated cells of the y-1 mutant of Chlamydomonas reinhardi, a unicellular green alga, a large amount of thylakoid membranes are formed in the chloroplast. However, the two major polypeptides of the membrane, which are required for formation of the structure of the membrane, are synthesized in the cytoplasm of these cells. One of these major polypeptides accumulated in the cell as a soluble protein when chlorophyll synthesis and membrane formation were inhibited.

This protein offers an opportunity to study directly the mechanism of membrane assembly. Experiments will be designed to follow in vivo the polypeptide from its existence as a soluble protein to its entry into the membrane. Standard procedures for purifying soluble proteins will be applied to this protein, and the interaction of the purified protein with chlorophyll (or soluble derivatives of chlorophyll) will be investigated.

Experiments will also be conducted to determine the mechanism by which light regulates the synthesis of the major membrane polypeptides in the cytoplasm as well as protein synthesis in the chloroplast.

SUPPORTED BY:
U.S. Natl. Science Foundation

CONTROL OF DNA METABOLISM IN CHLAMYDOMONAS,

S.H. HOWELL, Univ. of California, Graduate School, San Diego, California 92038

The DNA metabolic processes in the synchronous mitotic and meiotic division cycles of Chlamydomonas reinhardi are being studied by isolating conditional mutants with lesions in mitotic function. A subset of mutants is being selected which affect mitotic DNA metabolic processes, such as DNA synthesis and UV repair. These mutants are being screened for the amount of DNA synthesis, DNA breakage, and rejoining enzyme activities. The activity of the characterized conditional mutants is being tested under restrictive conditions of known cyclic DNA metabolic functions, such as DNA synthesis, in synchronous mitosis and meiosis. Special attention is being given to effects of these conditional mutants on genetic recombination and the formation of bi- parental recombinant molecules.

SUPPORTED BY:
U.S. Natl. Science Foundation

ENZYME TURNOVER DURING SYNCHRONIZED DEVELOPMENT,

R.F. JONES, State University of New York, School of Arts, Stony Brook, New York 11790

The long term objective of this research proposal is to examine the role that Catheptic proteases play in the turnover of specific enzyme substrates during synchronous development of the eukaryote Chlamydomonas reinhardi. In so doing the relative and specific roles that enzyme synthesis and enzyme degradation play in development and cell differentiation are being ascertained. The rate of de novo enzyme synthesis and the rate of enzyme degradation is being studied in order to more fully understand the molecular mechanisms of temporal and regulatory control governing enzyme accumulation during development and differentiation of young cells to their adult form.

SUPPORTED BY:
U.S. Natl. Science Foundation

ENVIRONMENTAL EFFECTS ON GROWTH AND PHOTOSYNTHESIS,

E.P. KARLANDER, Univ. of Maryland, Agricultural Experiment Sta., College Park, Maryland 20740

Objective: Study the effects of various light temperature interactions on the growth and photosynthesis of plants. Isolate and study the components of plant photochemical systems. Study the influence of various environmental factors, such as noxious gases, on growh, photosynthesis and metabolism of plants.

Approach: Culture plants at various temperatures in nutrient medium under different qualities of illumination adjusted to an equally- absorbed quantum flux. The growth, photosynthesis and certain components of the plants will then be measured. Extract the protein- pigment complexes and characterize them by biophysical procedures. When this is accomplished the stress of various environments on the plant can be assayed as to its particular effect on each system. Culture plants on nutrient media and employing several different atmospheres. Growth, photosynthesis and other indicated metabolic processes will be observed under each environmental condition.

Progress: NaCl toxicity in Chlorella sorokiniana resulted in decreased growth rates, increased dry weight per cell, increased intracellular Na and Cl, more total chlorophyll per cell, a decreased chlorophyll (a over -) to chlorophyll (b over -) ratio, increased rates of oxygen evolution, and decreased rates of carbon dioxide fixation. Inclusion of Ca ion in the medium extended the NaC1-tolerance of C. sorokiniana. Requirements for light-energy were higher in the NaC1- inhibited cultures.

SUPPORTED BY:
Maryland State Government - Annapolis

CELL DIFFERENTIATION DURING GAMETOGENESIS,
R.P. LEVINE, Harvard University, School of Arts, Cambridge, Massachusetts 02138

Gametic differentiation will be studied in Chlamydomonas reinhardi. This includes (1) a study of the metabolic changes that are first elicited by nitrogen starvation in the two mating types; (2) a study of possible changes in nucleic acid biosynthesis and nuclear proteins during gametogenesis; (3) a study of the transcriptional and translational requirements for gametogenesis; and (4) a study of the nature of the mating-type substances and their interaction at the time of mating.

SUPPORTED BY:
U.S. Dept. of Hlth. Ed. & Wel. - N.I.H.

AN INVESTIGATION OF PHOTOSYNTHETIC ELECTRON TRANSPORT WITH MUTANT STRAINS OF CHLAMYDOMONAS REINHARDI,
R.P. LEVINE, Harvard University, School of Arts, Cambridge, Massachusetts 02138

Mutant strains of Chlamydomonas reinhardi have been used successfully to study a number of different aspects of the photosynthetic processes. In this research they will be used in the analysis of two chloroplast-membrane associated functions: the photolysis of water and the transfer of excitation energy. In addition, their use will be extended to a new avenue of research, that of the molecular structure of chloroplast membranes. Several mutant strains are available for this study in which chloroplast membrane organization is altered. Their membranes will be analyzed biochemically and by techniques of molecular spectroscopy and compared with normal membranes.

SUPPORTED BY:
U.S. Natl. Science Foundation

THE GENETIC CONTROL OF CHLOROPLAST STRUCTURE AND FUNCTION IN CHLAMYDOMONAS REINHARDI,
R.P. LEVINE, Harvard University, School of Arts, Cambridge, Massachusetts 02138

This award involves continuation of the support provided under grant GB-18666 for Dr. Levine's studies on the genetic control of chloroplast structure and function in Chlamydomonas. This research is focused on the extent to which the structure and the function of an organelle are determined by genetic information contained organelle DNA and nuclear DNA, and the degree to which organelle ribosomes and cytoplasmic ribosomes participate in the translation of this information. The significance of this research relates to the fact that the biology of a eukaryotic cell is bound to the genetics, structure and function of its major cell organelles such as chloroplasts and mitochondria. These organelles and the cells in which they are found follow controlled growth conditions, under the direction of Dr. Michael Wynne, to test limits of variation in regard to such morphological criteria as cortication, trichoblasts, etc.

SUPPORTED BY:
U.S. Natl. Science Foundation

STUDIES IN PHOTOSYNTHESIS UNDER SALINE CONDITIONS,
R.N. NIEMAN, Shivaji University, Kolhapur, India

Objective: Study salt tolerance of plants with special reference to the process of photosynthesis.

Approach: Laboratory work is underway on "saline" varieties of rice. Another companion experiment is in progress on ferns. The investigation will study marine algae and higher plants growing in saline and brackish waters and study the process of photosynthesis as affected by saline environment. Particular attention will be given to the influence of salinity on carbon pathways and the basis for salt- induced changes in the products of photosynthesis.

Progress: Measured the photosynthetic activity of several marine algae. Studied in some detail the influence of NaCl on chlorophyll synthesis and on the rate of photosynthetic CO(2) fixation by Enteromorpha tubulosa. This is an intertidal marine alga that grows best in delta areas where the salinity of sea water is reduced by the outflow of fresh water. Its sensitivity to salinity makes it an interesting test alga. Found a positive correlation between the NaCl concentration of sea water, from 0.016 M to 0.6 M, and the chlorophyll content of E. tubulosa. On the other hand, maximum rates of CO_2 fixation were observed at low salinity (0.063 M NaCl). Thus, the enzymes limiting photosynthetic assimilation by this alga appear to function more efficiently at NaCl concentrations far below the usual concentration in sea water that produced the maximum chlorophyll content. Preliminary C labelling studies indicate that both C and C(6) photosynthetic pathways are operative in the several species of marine algae examined. Techniques have been devised for growing mangrove plants in the laboratory where salinity levels can be controlled. The influence of salinity on photosynthesis by these plants will be studied.

SUPPORTED BY:
U.S. Dept. of Agriculture - A.R.S. - S.W.C

TAXONOMY, ULTRASTRUCTURE, AND BEHAVIOR OF MARINE PRASINOPHYCEAE,

R.E. NORRIS, Univ. of Washington, School of Arts, Seattle, Washington 98105

A taxonomic, morphological, and biochemical study of marine Prasinophyceae on the Pacific coasts of North America and Japan will be conducted by cooperating scientists from the University of Washington, Tokyo Kyoiku University and Toho University. The research will focus on: (1) the identification and handling of cultures and field collections, and their behavioral responses; (2) an ultrastructural analysis of, particularly (a) the development of scales and their deposition on the cell membrane, (b) mitosis and cell division, (c) microtubules, peroxisomes, flagellar rootlets and basal bodies, and lysosomes and vacuoles, (3) an analysis of chromosome numbers and genome structure for as many different taxa as possible; and (4) the metabolism of mannitol and carbohydrate reserve substances.

SUPPORTED BY:
U.S. Natl. Science Foundation

NUTRITIONAL RELATIONSHIPS AMONG MARINE ORGANISMS,

L. PROVASOLI, Haskins Laboratories Inc., New Haven, Connecticut 06510

The previous research of Provasoli and others indicate that productivity in the food web is affected not only by levels and types of "classical" nutrients, but it is also dependent upon the release of metabolites and "special factors" during life and decomposition of many marine organisms. These chemical interactions between species seem far more extensive and meaningful than previously suspected.

After seaweeds have been purified from their microflora, certain species lose their typical morphology. Surprisingly, these are among the commonest green seaweeds, i.e. sea lettuce (Ulva lactuca) and its cousin Monostroma, which is widely cultivated as a food additive in Japan. Chemical purification of the filtrate factor for Monostroma is underway. Algal-bacterial cultures, microscopic analysis of the symbiosis, and bacterial cell disruption and fractionation are planned for Ulva and Polysiphonia.

The identification of the algal excretions has disclosed the release of a variety of biologically active compounds.

These advances permit a deeper look at the vitamin cycle, expecially of vitamin B-12. A survey of ecologically important algae will be initiated to confirm that several of them produce a B-12 binding factor.

The fresh-water crustacean Moina will be used to define the common basic nutritional requirements for filter feeding crustacea. This knowledge will be employed to define the more complex requirements of Artemia and ecologically important marine zooplankters. The Vitamin D requirement offers the hope of an eventual replacement of the now standard rat and chick assay.

SUPPORTED BY:
U.S. Natl. Science Foundation

CONTROL OF REPLICATION OF CYTOPLASMIC DNAS IN CHLAMYDOMONAS,

R. SAGER, City University of New York, Graduate School, New York, New York 10021

The specific aim of this research is to investigate the control of replication of organelle DNAs. Chlamydomonas provides a unique opportunity to study this problem both because nuclear and chloroplast DNAs can be readily separated in cesium chloride equilibrium density gradients and because of our extensive studies of cytoplasmic genetics with this organism. We now plan to use both genetic and biophysical approaches to correlate the physical and genetic behavior of organelle DNA. Replication of chloroplast DNA is being studied by following the incorporation of radioisotopic adenine of 15N-NH4 ion into total cell DNA, which is then extracted and fractionated in cesium chloride density gradients. We are examining the time and extent of replication of chloroplast DNA in cells grown under different physiological regimes, and examining the ability of various growth inhibitors to uncouple the replication of chloroplast and of nuclear DNAs. Detailed studies of the fate of the chloroplast DNAs from male and female parents, correlating physical and genetic data have been in progress for several years and are continuing.

SUPPORTED BY:
Amer. Cancer Society Inc. - New York, N.Y.

SULFUR METABOLISM IN ALGAE,

J.A. SCHIFF, Brandeis University, School of Arts, Waltham, Massachusetts 02154

Studies of sulfate reduction in intact cells of Chlorella, mutants of this organism, and cell-free systems obtained from it will be continued. Since the series of enzymatic steps leading from sulfate through adenosine 3' phosphate 5' phosphosulfate (PAPS) to thiosulfate have been defined and two enzymes necessary for the PAPS to thiosulfate conversion have been partially purified, the mechanism of this set of reactions will be explored further. Of particular interest is the source of the reduced sulfur moiety which is contributed by the enzyme extract. A series of mutants of Chlorella which are blocked for sulfate reduction have been obtained and these have been characterized with respect to their enzymatic constitution in thiosulfate formation and their abilities to reduce sulfate and other intermediates in vivo. These mutants will be explored further with a view to obtaining a clearer picture of the normal pathway of assimilatory sulfate reduction.

SUPPORTED BY:
U.S. Natl. Science Foundation

REGULATION OF ENZYME SYNTHESIS AND DEGRADATION, AND GENE EXPRESSION DURING THE PLANT CELL CYCLE,

R.R. SCHMIDT, Virginia Polytechnic Institute, School of Agriculture, Blacksburg, Virginia 24061

Objective: Study the biochemical mechanisms regulating the timing of expression of structural genes coding for enzymes catalyzing steps in nutrient assimilation (i.e., carbon dioxide fixation, nitrate and nitrite reduction, assimilation of exogenous ammonium and urea) and in ribonucleotide and deoxyribonucleotide biosynthesis, and to determine the mechanisms regulating the synthesis and degradation of these enzymes during the cell cycles of lower and higher eucaryotic plant cells. The intracellular location and site of synthesis of these enzymes and their genes will also be elucidated.

Approach: Synchronous cultures of both lower and higher plant cells in suspension culture will be prepared by a new isopycnic density- gradient technique. The biochemistry of the synchronous cell cycle will be approached using genetic analysis, subcellular organelle isolation, differential inhibitors of organelle and cytoplasmic enzyme synthesis, specific inhibitors of gene replication and expression, and enzyme fractionation and purification.

Progress: The structural genes of 2 inducible enzymes, isocitrate lyase (ICLase), TPNH-specific glutamate dehydrogenase (TPN-GDH), and 5 biosynthetic enzymes, ribulose-1, 5-diphosphate carboxylase (Carbox), aspartate transcarbamylase (ATCase) dihydroorotase (DHOase) phosphoribosylglycinamide synthetase (PRG-Syn), DPHN-specific glutamate dehydrogenase (DPN-GDH), are continuously available for transcription throughout the cell cycle of synchronous Chlorella cells. Heretofore, continuous gene availability was thought to be a characteristic of only bacteria. Enzyme degradation in vivo plays a major role in regulating the levels of Carbox and ATCase, whereas, PRG-Syn and DHOase are stable in vivo. Cyclic fluctuations in an endogenous stabilizer, tentatively identified as uridine monophosphate (pathway feedback inhibitor), contributes to the periodic stability of ATCase during the cell cycle. This is the first evidence that a pathway endproduct regulates both the level and activity of a pyrimidine pathway enzyme. The timing of gene replication in the cell cycle influences the pattern of accumulation of only ATCase and PRG-Syn during the cell cycle.

SUPPORTED BY:
U.S. Dept. of Agriculture - C.S.R.S.

CHEMISTRY OF SEAWEED AND BACTERIAL INTERACTIONS,

A.I. SCOTT, Yale University, Graduate School, New Haven, Connecticut 06520

Monostroma is a seaweed extensively cultured in Japan where it is used as a tasty food complement. When grown in the absence of the factor inducing normal morphology, Monostroma grows poorly and as disorganized loose single cells. The factor is produced and released in seawater by a few bacteria and seaweeds which grow in the same environment. The addition of the factor has dramatic effects on Monostroma: normal growth rate and development resume, resulting in the orderly production of the normal expanded leaf-like plant.

The study of the Monostroma factor and other interactions between bacteria and seaweeds has received continued funding by the Biological Oceanography Program (present NSF grant GA-33480X). A bioassay and a bacterial source of the factor are available and preliminary part-time chemical work has been done by Professor Scott, and Professor Provasoli will supply the active bacterial filtrates and do the bioassay to guide the fractionation.

Since the known plant hormones and other plant regulators are inactive on Monostroma, the Monostroma factor might be a new substance. Because it apparently stimulates growth, production of cell-binding polymers and regulates orderly organized growth, the understanding of its mode of action may transcend seaweeds.

SUPPORTED BY:
U.S. Natl. Science Foundation

ION TRANSPORT IN PLANTS,

R.M. SPANSWICK, State University of New York, School of Agriculture, Ithaca, New York 14850

The objective of the proposed research is to establish a quantitative description of ion transport in plants. The methods used will include radioactive tracer techniques for the measurement of ion fluxes and microelectrode techniques for the measurement of membrane potentials and resistances.

Internodal cells of the Characeae will

be used to test an hypothesis in which the electrical properties of the membrane are controlled by a voltage- dependent electrogenic H ion pump. Attempts will be made to measure pH of the cytoplasm under conditions where its value may be predicted from the hypothesis. Conditions have been established under which it should be possible to measure the active H plus efflux, which has already been estimated from electrical measurements, by measuring changes in external pH. Measurements of the ATP content of the cytoplasm will be made under a variety of conditions to determine whether ATP is the substrate for the H ion pump. Smith's hypothesis, in which the Cl ion influx in Chara corallina is driven by the pH gradient set up by the H ion pump, will be tested using Nitella translucens. Should the Cl ion influx in Nitella also depend on the pH gradient, a comparison of the energy sources for the H plus pump in the two species will be useful since Cl ion transport appears to be ATP- dependent in Chara but not in Nitella. The application of these hypotheses to higher plants will be tested using Elodea canadensis and Zea mays.

The electrophysiological methods used to demonstrate the existence of intercellular connections in higher plants will be used to investigate the intercellular connections in corn roots to establish the pathways of ion transport across the root. A test of the effectiveness of the Casparian Strip as a diffusion barrier and an attempt to identify the xylem vessels participating in exudation in the partially mature root will be made.

SUPPORTED BY:
U.S. Natl. Science Foundation

STUDY OF COPPER USED AS AN ALGAECIDE,

J.A. SWADER, Univ. of California, Agricultural Experiment Sta., Riverside, California 92502

Objective: Elucidate the mode of action of copper used as an algaecide and to determine the presence and effect of the ion content in the water of southern California recreational lakes on the solubility of the added copper. These waters have an alkaline pH and a rather high concentration of carbonate ions, an ideal situation for the formation of an insoluble precipitate of the copper and carbonate ions upon the addition of copper. Copper has to be applied at rates several fold higher than normal to obtain algal control. This induces an economical factor due to the higher rates of copper required as well as an environmental problem, with insoluble copper settling to the bottom. The use of chelating agents such as the diamine and triethanolamine salts may alleviate these problems by keeping the copper in solution. This would enhance the natural beauty and recreational use of these waters as well as relieving a potential pollution problem.

Approach: The effect of copper on the growth, metabolic activity, and ultrastructure of Chlorella, a unicellular algae, will be studied in an attempt to find the site of the copper inhibition. The ion content of lakes and ponds in southern California will be determined and the effects of these ions and chelating agents on the solubility and toxicity of copper will be observed.

SUPPORTED BY:
U.S. Dept. of Agriculture - C.S.R.S.

PHOTOSYNTHETIC PROCESSES IN CHLAMYDOMONAS,
R.K. TOGASAKI, Indiana University, School of Arts, Bloomington, Indiana 47401

Three aspects of regulation of photosynthesis in vivo will be studied using mutant strains defective in the photosynthetic apparatus. To facilitate the mutational analysis of photosynthetic mechanisms, new and efficient selection procedures for photosynthetic mutants will be developed. This will increase the efficiency of searching for new mutants and make easier the maintenance of mutant stock cultures. Also some positive selection procedure will allow one to enrich specifically for a certain class of mutants.

SUPPORTED BY:
U.S. Natl. Science Foundation

EFFECTS OF PCB'S AND SELECTED HERBICIDES ON BIOLOGY AND GROWTH OF ALGAE,
R.L. VADAS, Univ. of Maine, School of Arts, Orono, Maine 04473

The widespread nature of PCB's in the environment causes concern about effects on organisms forming the base of life-support pyramids (food and oxygen)--such as the algae. Specifically, concern focuses on: the possible inhibition of phytosynthesis and deleterious effects on ultrastructure. This project would determine: (1) growth rates and carrying capacities (in terms of resistance) of Platymonas subcardiformis, and other algae, at a range of concentrations of PCB's and other herbicides; (2) photosynthetic and respiration rates for Platymonas as a response to herbicides; (3) possible changes in cell ultrastructure with known inhibitory levels; (4) accumulation of herbicides by species, at sub-lethal and lethal concentrations.

In the procedure planned: Growth rates will be determined by celloscope counting of inoculated stock cultures of algae, in which runs are terminated at herbicide carrying capacity. Photosynthetic response to herbicides will be determined by placement of the algae in a lighted Gilson differential respirometer, following various treatments at phases of light-dark cycle. Periodically cells will be sectioned and the ultrastructure observed under the electron microscope. PCB and DDT accumulations will be determined by gas chromatography.

SUPPORTED BY:
U.S. Dept. of Interior - O. Wtr. Res. Rch.

ANTI-ALGAE COMPOUNDS,
R.T. VANALLER, Univ. of Southern Mississippi, School of Science, Hattiesburg, Mississippi 39401

Isolation of and structural studies on a naturally occurring inhibitor of algae. The material inhibits growth of green and blue- green algae commonly found in pollution.

SUPPORTED BY:
University of Southern Miss. - Hattiesburg

ESTUARINE PRODUCTIVITY,
G.E. WALSH, U.S. Environ. Protection Agcy., Gulf Breeze Environ. Res. Lab., Gulf Breeze, Florida 32561

Pesticides and other substances are being tested for their effects on marine unicellular algae (16 species), seagrasses (Thalassia testudinum and Ruppia maritima), and red mangrove (Rhizophora mangle) seedlings.

Effects of the substances on growth and photosynthesis by algae and on survival and translocation in seagrasses and mangrove seedlings are studied.

Unicellular chlorophytes, chrysophytes, and rhodophytes in our collection were generally more susceptible to urea and triazine herbicides than were bacillariophytes. Survival of T. testudinum over a 2-week period was not affected by the herbicides 2,4-D, Picloram, Chloramben, Atrazine, Ametryne, Diclobenil, Diquat, Paraquat, Silvex, or Trifluralin. Mangrove seedlings were killed by a combination of 4.4 kg/ha 2,4-D and 1.6 kg/ha Picloram applied to the soil.

SUPPORTED BY:
U.S. Environ. Protect. Agency - O.R.M.